AF327520

❧ Foreword ❧

From the inception of the Hunt Institute's program in botanical art and illustration, we have laid particular emphasis on artists and works of the twentieth century. Of course, the crowning glory of our collection will always be its original nucleus—a splendid assemblage of works dating mostly from 1500-1850, gathered over many years by Rachel McMasters Miller Hunt. At the founding of the Institute, however, there was little prospect for significant further growth of the collection in works of that earlier period; coincidentally, there was apparent a distinct need for active, comprehensive documentation and promotion of modern botanical art. Hence our choice of particular, though by no means exclusive, endeavor. This emphasis is reflected alike in our acquisitions, by now largely restricted to contemporary works, and in our exhibitions.

Like the earlier exhibitions in this continuing series on twentieth century botanical art and illustration, the present show bears a special relationship to the Institute's permanent collection. Although many of our exhibitions include modern artists who are represented in the collection, the full scope of our twentieth century holdings is apparent only in these International Exhibitions. Collectively, they account for nearly every modern artist whose work we hold. Also, in consequence of our sustained emphasis in acquisitions, the series provides a fairly broad view of the field as a whole. The associated set of catalogues thus constitutes an incremental, reasonably representative accounting of modern, and especially contemporary, botanical art and illustration.

The basic rationale of the series remains unchanged—each exhibition is focused on works added to the collection since the time of the previous one. However, though the original plan was to exhibit only works drawn from our own collection, we have begun including some borrowed works as well. In each exhibition, this expanded compass permits more faithful reflection of the current scene; it also affords greater variety over the series as a whole.

We are aware of quite a few artists still unrepresented in our collection and in these exhibitions. Over the next few years we hope to acquire works by these and many others as yet unknown to us, particularly from those countries so far underrepresented or, worse, unrepresented. Accordingly, we invite all such artists and illustrators whose work has been (or will be) published in some form to make themselves known to us. We also solicit referrals to such artists from others who may know of their work. One of our chief functions is the preservation, in a muscum setting, of original art-

Entrance to the Main Exhibition Gallery

works for the future, particularly those originals for published illustrations which are often otherwise lost.

This Fourth International Exhibition testifies to the continued variety, vigor, and beauty of contemporary botanical art. The times are evidently quite propitious for utilization, appreciation, and advancement of this time-honored genre; the Hunt Institute is pleased to have a part in its support and to offer this exhibition and catalogue in its celebration.

Robert W. Kiger
Acting Director

International Standard Book Number: 0-913196-19-3

Catalogue

4th International Exhibition of Botanical Art & Illustration

6 November 1977-31 March 1978

❧

With an Introduction by John V. Brindle

Compiled by Sally W. Secrist and N. Ann Howard

❧

The Hunt Institute for Botanical Documentation

Carnegie-Mellon University, Pittsburgh, Pennsylvania 1977

❧ Introduction ❧

The appeal of botanical art is both aesthetic and scientific. /Organic patterns of growth and rhythm inherent in plant forms reveal a harmony and balance to which artists have always responded, and plant forms have been a major theme of art through the ages. Our age particularly finds direct aesthetic satisfaction in the geometry of nature's organic patterns, a trait perhaps resulting from the conditioning effect of prolonged exposure to abstract art. /For the science of botany, illustration has had a major role. From the days when interest in plants was confined to their curative properties described in herbals decorated with crude figures, and throughout the development of botany as a true science, progress has depended heavily on the wide dissemination of printed books, to whose texts illustration has been essential.

This fourth in our series of international exhibitions of twentieth century botanical art and illustration offers an occasion to assess the contemporary state of affairs. The artworks making up the exhibition (and the Hunt Institute collection from which most have been drawn) are widely varied, but viewed *in toto* they document a vigorous recovery from a low state that prevailed from, roughly, mid-nineteenth century well into the twentieth— a decline brought about by the stultifying sentimentalism of the time, by uncritical adoption of novel color-printing processes, and by the challenge of photography. Today's botanical artist enjoys a market of greatly expanded demand and a more sophisticated appreciation of his works. At his disposal is an array of media and manners which he exploits at a general level of competence that does honor to his craft, and the images he produces are distributed on an unprecedented scale.

The exhibition reflects a continuation of the close relationship to publication that prevailed in earlier periods. More than one-third of the artworks displayed are originals for book illustrations of some kind, ranging from drawings for textbooks or botanical monographs to paintings for color plates of garden favorites, published for the same market that made the Florilegium popular in seventeenth century Europe. Other entries represent different modes of publication. Thus Manabu Saito's "Maidenhair Fern" has been issued as an independent color reproduction; Dr. Otto Ludwig Kunz's "Oriental Poppy" illustrates a calendar; Dagny Tande Lid's sketches are designs for postage stamps, and Ethel Dean's for fabrics. There are several examples of "art prints" (hand-pulled prints, signed by the artist), also a form of publication. Among these are Luigi Lucioni's etchings— landscapes in which trees are the main features, and which may also be seen as habitat groups of birches, pines, and maples—and Henry Evans' continuing series of subtly designed linocuts.

Today's artist enjoys a nearly unrestricted freedom of choice in selecting a mode and medium of expression, and it is interesting to note how many have adopted (and adapted) traditional techniques that had originally been developed simply as economical processes for multiple replication. Woodcut, a relief process readily incorporated into the letterpress of the early years of printing, was the vehicle of choice for Mervin Jules' "Sunflowers." Etching, which offered new flexibility and refinement to seventeenth century illustrators and publishers, was taken up by Marvin Spohn to convey his often whimsical visual comments. Lithography, which afforded the illustrator a less laborious means of reproduction, and greatly expanded the quantity of illustrations in the nineteenth century, is favored by Kenjilo Nanao to realize his special vision. In such instances a medium is chosen for its intrinsic character, to achieve an aesthetic effect that no other process could precisely match.

Among today's artists, watercolor remains the favorite medium (as it has been since the eighteenth century) to capture the fragile vitality of the flower. In this medium, the exhibition offers an embarrassment of riches. Among others, Rory McEwen of England, Yoai Ohta of Japan, Marilena Pistoia of Italy, Elisabeth Schroer of Switzerland, František Severa of Czechoslovakia, Elsa Felsko and Marianne Golte-Bechtle of Germany, and Anne Ophelia Dowden of the United States have mastered this "not-as-easy-as-it-looks" technique. Color illustrations, reproduced by various modern photomechanical processes, enliven all sorts of botanical publications, from textbooks to calendars, thus supplanting the traditional dependence of handcoloring (with watercolor washes) woodcut, engraved, or lithographed plates individually. Admittedly, yet another element of antique charm is thus yielded to progress, but there is gain in making attractively illustrated books available to a market well beyond anything that an eighteenth century publisher could imagine.

Production cost problems have been, from the beginning, as dear to the publisher's heart as taxes. Important botanical texts have been printed unadorned when the author or publisher could not afford the luxury of engraved figures, and budgets are still often restrictive. In such cases ink drawings are the answer almost universally, being reproducible at minimal cost as "line-cuts." Fine examples of ink drawing are too abundant in the exhibition for individual citation. The range is from simple outline plant-identification figures to studies whose forms are fully rendered through the tonal technique of stipple (see Cleuter drawings). Though drawing is a comparatively simple and limiting medium, its use does not deprive the illustrator of the opportunity for creativity, but merely poses a sharper challenge.

Collectively, the catalogues of the four exhibitions in this series, begun in 1964, account for more than 400 artists and illustrators, a respectable number surely, but many qualified artists are still unrecorded, and the list cannot be termed comprehensive. It may be regarded however as fairly representative of work being done in this specialized field and justifies a few general remarks.

As was the case in the three earlier exhibitions (and, indeed throughout the history of botany), this one includes several entries by botanists who illustrate their own texts: Dr. Tetsuo Koyama of the New York Botanical Garden draws with a distinctive style reflecting his oriental heritage; the boldly decorative drawings of Dr. Geoffrey Herklots enliven his books; Esmé Hennessy of South Africa, whose entries surely indicate the capacity for a successful career as an artist-illustrator, distinguishes her works as either "scientific" or "decorative" and signs them differently. Knowledge of plant structure, careful observation, and respect for accuracy are essentials of a botantist's training, but in these instances, there is also apparent the artist's intuitive grasp of graphic expression.

A high percentage of women in the list—55% of the names in all four catalogues, and nearly 59% in the current exhibition—is not unexpected; it merely reflects a long-standing trend. This fourth international includes a greater than usual proportion of Americans, attributable to a purchasing grant from the National Endowment for the Arts during the Bicentennial year. Many of the artists and illustrators working in America, however, were born and trained outside the United States. Apparently artists still gravitate toward the liveliest-looking market. In this connection, in any case, it would be difficult to demonstrate the existence of national schools of botanical art, though a systematic study based on a truly comprehensive scope might modify the statement.

A noteworthy aspect of this exhibition is the commingling of art and illustration. There is no attempt here to draw a precise boundary between the two, though the observer may choose to do so for himself. If so, he will have occasion to note a strong aesthetic element in many of the illustrations; and on the other hand to observe that many of those works conceived as art are more than adequately descriptive. Thus, in Karin Douthit's ink illustrations, which satisfy the botanist's exacting criteria, one is struck by a beautifully controlled tonality and a sensitive disposition of parts on the sheet that qualify them as works of art, while Rory McEwen's elegant watercolor portraits on vellum for *Tulips and Tulipomania* would satisfy the keenest botanical eye. If the distinction between artist and illustrator is thus often blurred, the resulting mélange may be regarded as a happy state in which both art and science are well served.

Many of the artworks in this exhibition have been donated to the Hunt Institute, or lent for the occasion, as indicated in the catalogue listing.

The National Endowment for the Arts has made it possible to acquire the works of American artists through a purchasing grant.

Dr. H.C.D. de Wit, of the Laboratorium voor Plantensystematiek en -Geografie, Landbouwhogeschool, at Wageningen, Netherlands, contributed generously of his time and effort in gathering original drawings by illustrators connected with his institution, and arranging for their deposit at the Hunt Institute as a permanent loan. Dr. de Wit also compiled biographical data on the illustrators and furnished their photo-portraits.

Mrs. Gertrude Espenscheid of the Hunterdon Art Center at Clinton, New Jersey has been helpful in leading us to several new artists, and in arranging for the transfer here of artworks after the closing of the Center's summer exhibition.

Mr. Samuel Howard, Art Director at *Scientific American,* helped us in acquiring artworks from Mr. Bunji Tagawa and Mr. Thomas Prentiss.

Dr. and Mrs. Tom Fujii, of Tokyo, freely offered their services in corresponding with the Japanese artist, Yoai Ohta.

Mr. Sylvan Cole, Jr. and Mrs. Meg Dunwoody, of Associated American Artists, New York, and Mr. Arnold Klein, of Arnold Klein Gallery in Royal Oak, Michigan, put us in touch with several new artists and sent artworks for our selection.

To all these benefactors and to others who have been helpful, I wish to express, on behalf of the Hunt Institute, our deep gratitude for the generosity and cooperation which have so greatly enriched this exhibition and the collection as a whole.

Among the Institute staff, Drs. Kiger, Daniels, and Buchheim all helped with problems of plant identification and nomenclature; Mrs. Sally Secrist (designer of this catalogue) and Mrs. Ann Howard did much of the necessary correspondence and worked with skill and cheerful patience at all stages in the preparation of both exhibition and catalogue; Mr. Joseph A. Rosen photographed artworks for the catalogue; Mr. Joseph Calcutta, Miss Linda Deak, Miss Claire Wallace, and Mr. John Golec all did essential work in the preparation and mounting of the exhibition; Mrs. Karen Britz and Mrs. Donna Connelly typed the correspondence and the catalogue copy; Mrs. Rita Gordon and Mrs. Estelle Weissburg bore the burden of the often-complicated bookkeeping operations. I thank them all.

John V. Brindle
Art Curator

❧ Notes to the Exhibition ❧

Works in this exhibition are numbered sequentially following the alphabetical listing of the artists in the catalogue.

The common name of the plant, or the title of the work as given by the artist, is followed by the Latin name of the plant in italics. Where a Latin name has been supplied by the artist, no attempt has been made to provide synonyms.

Dimensions cited (in inches, height first) are for the mat openings, unless otherwise noted.

Exhibition items known to be the originals used in a published work are so indicated by an asterisk after the appropriate publication in the artist's bibliographic entry.

All items in the exhibition, unless otherwise noted, are from the permanent collection of the Hunt Institute.

ABBE, Elfriede

Born   Washington, D.C.,
6 February 1919.

Address   Applewood, Manchester Center, Vermont 05255.

Education   Cornell University, College of Architecture: B.F.A., 1940. Art Institute of Chicago: 1937.

Career   Graphic artist, sculptor, typographer, and book designer. Formerly, Scientific Illustrator, Division of Biological Sciences, Cornell University, 1942-74.

Media   Wood engraving, wood sculpture, letterpress.

One-person Exhibitions   White Art Museum, Cornell University, 1963; Trinity College, Hartford, 1964; Southern Vermont Art Center, 1964, 1967, 1974; Pen and Brush, New York, 1965; Print Club of Albany, 1968; Hunt Institute, 1969; Arts Club of Washington, D.C., 1972.

Group Exhibitions   Numerous, including: National Academy; National Arts Club; National Sculpture Society; Salmagundi Club; London Chappel Exhibition, London, Stanbrook Abbey, Worcester and Birmingham, 1963; Hunt Institute Internationals, 1964, 1968, 1972; Hunt Institute Travel Shows (Botanical Prints, Plants in Art).

Honors/Awards   Numerous, including: Pen and Brush, New York, Gold Medal, 1964; Tiffany Fellowship, 1948; Roy Arthur Hunt Foundation grant, 1961; National Arts Club, Bronze Medal, 1966, 1967, Gold Medal, 1970; Pen and Brush, Founders' Prize, 1977.

Collections   Represented in more than 20 public collections, including: Museum of Fine Arts, Boston; Cincinnati Art Museum; Dumbarton Oaks, Washington, D.C.; Henry E. Huntington Library and Art Gallery, San Marino, California; Houghton Library, Harvard University; New York Botanical Garden; New York Public Library; St. Bride Institute Printing Library, London; Rosenwald Collection, National Gallery; many universities and colleges throughout the United States and Canada.

Commissions/Works Published in   From the Abbe Press: *Aesop's fables,* 1950; *Rip van Winkle,* 1951; *Prometheus bound,* 1952; *Garden spice and wild pot-herbs,* 1954; *The revelation of St. John the Divine,* 1958; *The Georgics of Virgil,* 1966; *Mushrooms* (portfolio), 1967; *The creation.*
From the Cornell University Press: *The significance of the frontier in American history,* 1956; *Seven Irish tales,* 1957; *The plants of Virgil's Georgics,* 1965; R.T. Clausen: *Sedums of the Trans-Mexican belt,* 1959.
Sculptures for Anne Arundel College; Cornell University; Emma Willard School, Troy, New York; McGill University; New York Botanical Garden; New York World's Fair, 1939; Tompkins County Public Library, Ithaca, New York; Unitarian Church, Ithaca; Vermont Council on the Arts.

1   "Fiddleheads"
Wood engraving; 6⅞ x 4⅝"

2   "Indian Pipes," *Monotropa uniflora*
Wood engraving; 8¼ x 5⅜"

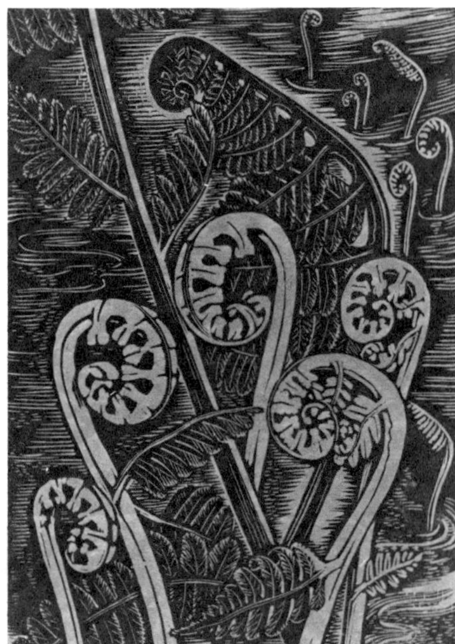

1

ALAVA, Reino Olavi

Born   Paimio, Finland, 21 May 1915.

Address   ℅ Department of Botany, University of Turku, SF-20500 Turku 50, Finland.

Education   University of Turku, Turku, Finland: M.A., Botany 1947. Washington University, St. Louis, Missouri: Ph.D., Botany, 1951. Self-taught botanical illustrator.

Career   Curator of the Herbarium, University of Turku, Turku, Finland, and botanist specializing in the taxonomy of Umbelliferae. Formerly, botanical illustrator, Department of Botany, University of California, Berkeley, 1955-58.

Medium   Ink.

Group Exhibition   Hunt Institute International, 1964.

Collections   Department of Botany, University of California, Berkeley; Department of Botany, University of Turku.

Works Published in   Alava, R. "Spikelet variation in *Zea mays*," *Annals of the Missouri Botanical Garden*, Vol. 39, 1952 (Ph.D. thesis).
Mathias, M.E. and Constance, L. *Bulletin of the Torrey Botanical Club*, Vol. 84, 1957; Vol. 85, 1958; Vol. 89, 1962; *University of California publications in botany*, Vol. 33, 1962; Vol. 38, 1965.
Alava, R. " The genus *Ainsworthia*," *Notes of the Royal Botanic Garden, Edinburgh*, Vol. 31, 1971.*
Alava, R. "A new genus of Umbelliferae (Tordyliinae) from Caucasia," *Notes of the Royal Botanic Garden, Edinburgh*, Vol. 32, No. 2.
Alava, R. "The genus *Ducrosia* and its allies," *Notes of the Royal Botanic Garden, Edinburgh*, Vol. 34, 1975.
Alava, R. *"Kandaharia*, a new genus of the Umbelliferae from southeast Afghanistan," *Candollea*, Vol. 31, 1976.

3 *Ainsworthia cordata* *
Ink; 16½ x 10¼"

4 *Ainsworthia trachycarpa* *
Ink; 16⅜ x 10"

Gifts of the artist

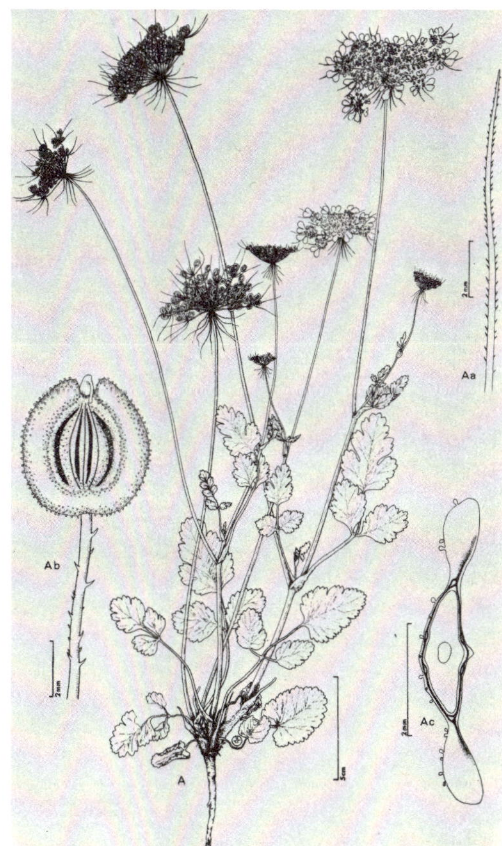

4

ALLEN, Dorothy Osdieck (Mrs. Paul H.)

Born   Kirkwood, Missouri, 9 May 1911.

Died   St. Louis, Missouri, 29 April 1973.

Education   Kirkwood High School, Kirkwood, Missouri.

Career   Self-taught botanical illustrator of papers and books by the late Paul H. Allen, 1947-63.

Medium   Ink.

Group Exhibition   Hunt Institute International, 1964.

Works Published in   Allen, Paul H. "Orchidaceae" in Woodson, R. E. and others. " Flora of Panama," *Annals of the Missouri Botanical Garden*, Vol. 36, 1948.*
Allen, Paul H. *The rain forests of Gulfo Dulce*. Gainsville, Florida, University of Florida Press, 1956 (Reissue by Stanford University Press in progress).**

5 Orchid, *Cattleya skinneri* *
 Ink; 14¼ x 9⅜"

6 Orchid, *Oncidium powelli* *
 Ink; 14¼ x 9⅜"

7 *Couratari panamensis***
 Ink; 13½ x 10⅛"

8 *Enterolobium cyclocarpum***
 Ink; 13½ x 11¼"

Gifts of the artist

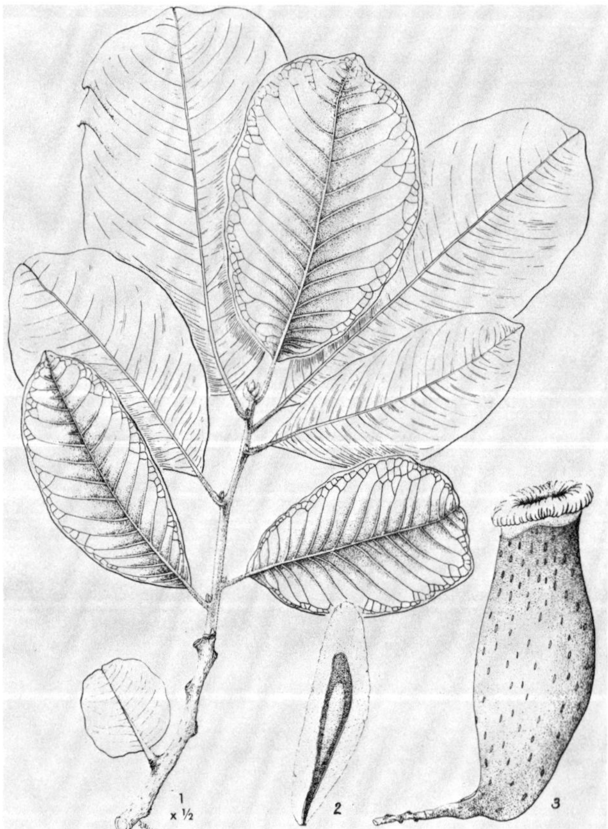

7

ANDREWS, Jennifer

Born   Oxford, England, 20 March 1933.

Address   Mill House, Henley, Ipswich, Suffolk, England.

Medium   Ink, watercolor.

Education   Oxford School of Art. Royal College of Art, London.

Career   Free-lance artist and designer.

One-person Exhibitions   Minories, Colchester, Essex; Mercury Theatre, Colchester, Essex.

Group Exhibitions   Bear Lane Gallery, Oxford; Deben Gallery, Woodbridge, Suffolk; Scottish Arts Council.

Collections   John Nash; Mary Grierson; Dr. and Mrs. R.B. Burbidge.

Work Published in   Burbidge, R.B. *A dictionary of British flower, fruit, and still life painters.* Lehigh-on-Sea, F. Lewis, Publishers, Ltd., 1974.

9 Bellflower, *Campanula lactiflora* 'Loddon Anna'
   Ink and watercolor; 19½ x 12¼″

10 Hellebore, *Helleborus orientalis*
   Ink and watercolor; 9 x 12⅞″

11 Narcissus
   Ink and watercolor; 13 x 10½″
   All lent by the artist

9

ANGEL, Marie Felicity

Born   London, England,
15 March 1923.

Address   Silver Ley, Oakley Road,
Warlingham, Surrey CR3 9BE,
England.

Education   Croydon School of Art
and Craft: 1940-45. Royal College
of Art, School of Design: Diploma
of Associateship, 1948.

Career   Free-lance illustrator and
calligrapher.

Medium   Watercolor on vellum.

One-person Exhibitions   Palace of the Legion of Honor, San
Francisco, 1967; Casa del Libro, Puerto Rico, 1974.

Group Exhibitions   Society of Scribes and Illuminators;
Society of Designer Craftsmen of Great Britain; Crafts
Centre of Great Britain; Royal Academy Summer Exhibition;
Crafts Council of Great Britain; Hunt Institute
Internationals, 1964, 1968, 1972; Hunt Institute Travel Shows
(International).

Honors   Member, Society of Scribes and Illuminators; Fellow,
Society of Designer Craftsmen of Great Britain.

Collections   Victoria and Albert Museum, London; Harvard
College Library, Cambridge, Massachusetts; Casa del Libro,
Puerto Rico; and many private collections throughout
England, Europe, and the United States.

Commissions/Works Published in   Angel, Marie. *A bestiary.*
Cambridge, Massachusetts, Harvard College, 1960.

Angel, Marie. *A new bestiary.* Cambridge, Massachusetts,
Harvard College, 1964.

Fisher, Aileen. *We went looking.* New York, Thomas Y.
Crowell, 1968.

Two poems by Emily Dickinson. Cambridge, Massachusetts,
Harvard College Library, and New York, Walker and
Company, 1968.

The twenty-third psalm. New York, Thomas Y. Crowell,
1970.

Potter, Beatrix. *The tale of the faithful dove.* New York
and London, Frederick Warne and Company, 1971.

An animated alphabet. Cambridge, Massachusetts, Harvard
College Library, 1971.

Potter, Beatrix. *The tale of tuppenny.* New York and
London, Frederick Warne and Company, 1973.

Fisher, Aileen. *My cat has eyes of sapphire blue.* New York,
Thomas Y. Crowell, 1973.

Angel, Marie. *The ark.* New York, Harper and Row, 1973.

Talbot, Toby. *Two by two.* Chicago, Follett Publishing
Company, 1974.

Beasts in heraldry. Brattleboro, Vermont, The Stephen
Greene Press, 1974.

Angel, Marie. *The art of calligraphy, a practical guide.*
New York, Charles Scribners' Sons, 1977.

Abecedarium series of plant subjects, commissioned by the
Hunt Institute, 1964 to the present [in progress]. *

12 Common Thyme, *Thymus vulgaris* *
Watercolor on vellum; $3\frac{5}{8}$ x $4\frac{1}{8}$"

13 Sweet Briar, *Rosa eglanteria* *
Watercolor on vellum; $3\frac{5}{16}$ x $3\frac{9}{16}$"

14 Marjoram, *Origanum vulgare* *
Watercolor on vellum; $3\frac{3}{8}$ x $4\frac{1}{16}$"

15 Heartsease, *Viola tricolor* *
Watercolor on vellum;
$3\frac{13}{16}$ x $3\frac{11}{16}$"

14

BALBI, Domenico

Born Genoa, Italy, 1 February 1927.

Address Via Burlando 12A/8, Genoa, Italy.

Education Academy of Fine Arts, Genoa: 1945-49.

Career Painter, designer, engraver.

Media Oil, intaglio.

One-person Exhibitions San Matteo, Genoa, 1951; Il Prisma and Vinciana, Milan, 1958; Ferrari, Verona, 1960; del Libraio, Bologna, 1961; Municipal Library, Milan, 1961, 1965; Nerea, Udine, 1961; Pater, Milan, 1965; dell'Arnetta, Gallarate, 1965; Il Salotto, Como, 1966; Il Camino, Rome, 1966; Vallombreuse, Biarritz, France, 1969; Mouffe, Paris, 1970; The Breakers, Palm Beach, Florida, 1972.

Group Exhibitions XV Premio Suzzera, 1962; Modern Art Centre Max G. Bollag, Zurich, 1965; VII Biennale d'Arte Sacra, Bologna, Milan, 1966; Biennale delle Regioni, Ancona, 1967-69; Biennial of the Flowers in Graphics, Pistoia, 1968; 2nd Annual of Graphic Art, International Exhibition, Ancona, 1968; Grand Prix de Peinture du Fest., Cannes, 1968; Contemporary European Painters, New York, 1968; International Autumn Salon, Biarritz, 1970; Award of Graphic Art "Free Meeting," Venice, 1971; 7th National Exhibition of Italian Engraving, Sassari, 1971; Corner Gallery, London, 1971; Art Basel, Cologne, 1973.

Honors/Awards Honorary Diploma, Biennial of Regional-European Art, Ancona, 1967; First Prize "Citta d'Imperia," 1966; Golden Trophy, Grand Prix of Painting, Festival of Cannes, 1968; First Prize for Still Life, Fifth Grand Prize, Painters of Summer, Nice, 1968; Diploma of International Notification, Second Annual International Exhibition of Graphic Art, 1968; Diploma of Honor, 3rd Biennial Regional Show, 1969; Europa Arts Tetradramma d'Oro, Corriere di Roma, 1974.

Collections Many public collections, including: Castello Sforzesco, Milan; National Museum, Pisa; Civic Museum, Udine; Municipal Libraries, Milan and Genoa; S.A.S. Prince of Monaco.

Works Published in *Tarot Balbi,* Vitoria (Spain), Heraclio Fournier, 1976.
Playing cards by Italsider, 1976.
Contributor to many books, pamphlets, and encyclopedias.

16 "Mixed Flowers in a Vase"
Engraving; 12⅛ x 18¾" plate mark

BARBISAN, Giovanni

Born Treviso, Italy, 6 April 1914.

Address Via Montepiana 12, Treviso, Italy.

Education Academy of Fine Arts, Venice: Diploma.

Career Artist and printmaker.

Medium Etching.

One-person Exhibitions Gallery of Modern Art, Venice, 1974; Braschi Palace, Rome, 1976.

Group Exhibitions Biennial Art Exhibitions, Venice, 1936-66; 5 Quadrennial Art Exhibitions, Rome; Exhibitions of Italian Art in Berlin, 1937, Cairo, 1959, Lisbon and Stockholm, 1953; major museums at Boston, Stockholm, Lima, Wiesbaden, Tokyo, Brussels, Moscow, etc.

Awards Biennial of Art, Venice, 1940, 1950; Exhibition of Fine Arts, Turin, 1953; Quadrennial of Rome, 1953; Premio Michetti, 1952; Bevilacqua La Masa, 1946; Premio Burano, 1954; Premio Marzotto, 1955; National Exhibition of Portraits, Venice, 1955; Reggio Emilia, 1957; La Spezia, 1957; National First Prize "A. Soffici," 1966; many others.

Collections Numerous, including: the modern art galleries of Venice, Rome, Turin, and Verona; Sforzesco Castle, Milan; Stockholm Museum; other collections in Brussels, Bern, Boston, Philadelphia, New York, Moscow, Cairo.

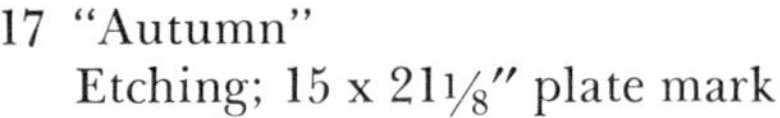

17 "Autumn"
Etching; 15 x 21⅛" plate mark

BARTEL, Maureen (Mrs. Herbert W.)

Born   Liverpool, England, 16 October 1933. Has worked in the United States since 1956.

Address   5401 Wellesley Avenue, North Olmsted, Ohio 44070.

Education   Self-taught botanical artist.

Career   Free-lance botanical artist. Teaches nature drawing at Forest Elementary School, North Olmsted, Ohio. Formerly, with the British Consulate-General, Cleveland.

Medium   Acrylic.

One-person Exhibitions   Shaker Lakes Regional Nature Center; Rocky River Trailside Interpretive Center; Baldwin Wallace College; Lake Erie Junior Science and Nature Center; Garden Center of Greater Cleveland, 1974; Holden Arboretum, 1975; Cleveland Public Library, 1975.

Works Published in   The American Nature Study Society publications.

18 Jewel-Weed, *Impatiens biflora* and *I. pallida*
Acrylic; 9¾ x 9⁵⁄₁₆″

19 *Rosa virginiana*
Acrylic; 10¼ x 8″

18

BATA-GILLOT, Farida Magdalena

Born   Djakarta, Indonesia,
26 December 1950.

Address   Thorbeckestraat 108,
Wageningen, Netherlands.

Education   Academy of Art,
Arnhem.

Career   Painter and graphic artist.
Botanical artist, State Agricultural
University, Wageningen, 1973 to
present.

Works Published in   Botanical
periodicals.

20 *Celosia argentea*
Ink; 11 x 6⅝″

21 *Acridocarpus spectabilis*
Ink; 10¼ x 7¹⁄₁₆

Permanent loans of the Laboratorium voor
Plantensystematiek en -Geografie,
Landbouwhogeschool, Wageningen, Netherlands

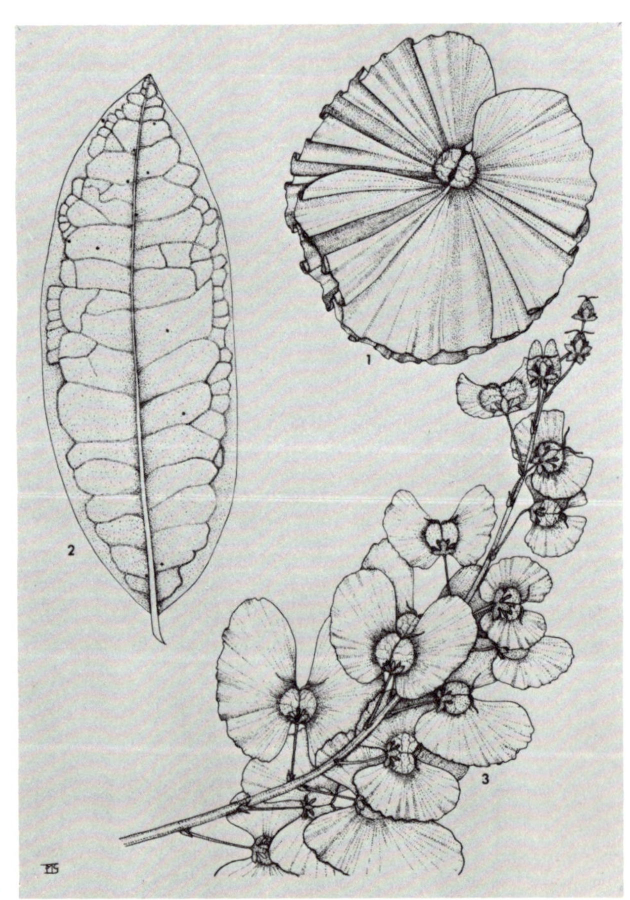

21

BAUER, Alma Gene (Mrs. Dale W.)

Born   Garden Grove, California, 16 October 1926.

Address   Box 205, Running Springs, California 92382.

Education   University of California, Los Angeles: B.A., 1953.

Career   Free-lance artist. Formerly, art teacher, Los Angeles City Schools, 1954-61.

Media   Serigraphy, ink, pencil, opaque watercolor.

One-person Exhibition   Gavilan College, Gilroy, California, 1976.

Collection   Helen Crocker Russell Library, Strybing Arboretum, San Francisco.

Works Published in   *Golden gardens;* San Bernardino *Sun-telegram; The national gardener.*
Bauer, G.   *The golden native.* California Garden Clubs, Inc., 1972-74.
Bauer, G.   *Golden botanical gardens.* California Garden Clubs, Inc., 1976-78.*

22 Fuchsia 'Trumpeter' *

23 Sweet Orange, *Citrus sinensis* *

24 Passion-Flower, *Passiflora jamesonii* *

25 Prickly-Pear, *Opuntia acanthocarpa* *

26 Camellia *

27 Tulip-Tree, *Liriodendron tulipifera* *
Serigraphs; 9 x 7⅛″ sheet sizes
Gifts of the artist

22

"

BECKETT, Sheilah Tannis

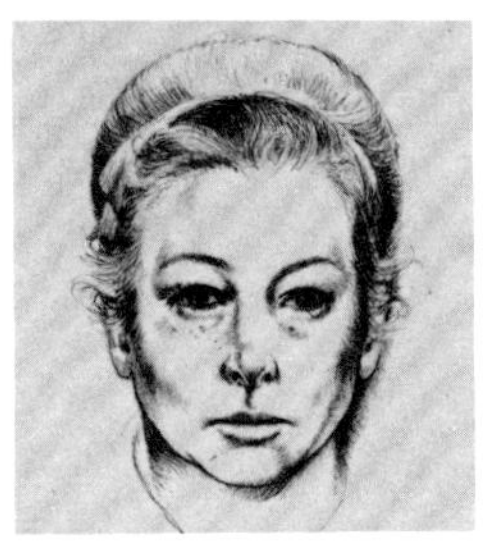

Born Vancouver, British Columbia, Canada, 5 September 1913. United States citizen at an early age.

Address "Eirinn," Ogden Road, Ossining, New York 10562.

Education Museum Art School, Portland, Oregon. Chouinard Institute of the Arts, Los Angeles. Studied painting with Anthony Tony and sculpture with Helen Belling and Hertzel Emanuel.

Career Free-lance commercial artist and illustrator, New York, 1940 to present. For many years, with Charles E. Cooper Studio, New York.

Media Watercolor, sculpture (in clay, lead, bronze, papier-maché).

Group Exhibitions Art of the Nativity, Tarrytown, New York (3 years).

Collections Nathaniel Hawthorne College, New Hampshire; sculpture in private collections.

Commissions/Works Published in *Esquire, Coronet,* and *Jack and Jill.*
Over thirty books (mostly for children) published by Bantam Books, Dell Publications; Grosset and Dunlap; Little, Brown; Random House; Simon and Schuster; Jonathan Cape.
Christmas cards for American Artists Group, New York (25 years).

28 "Long Ago, Last Summer"
Pencil and watercolor; 7⅝ x 10⅞"

29 "Ophelia's Garland"
Pencil; 12 x 10½"

28

BERO, Robert

Born   New York, New York, 6 June 1941.

Address   Fox Hill Road, Tuxedo Park, New York 10987.

Education   Pratt Institute, New York: B.F.A., 1964. Yale University, New Haven, Connecticut: M.F.A., 1966.

Career   Painting and printmaking instructor, Ramapo College, Mahwah, New Jersey, 1976 to present. Formerly, printmaking instructor State University of New York, Pottsdam, 1966-73. Taught at Pratt Institute, New York, and Yale University. Printmaking instructor, Brown University, Providence, Rhode Island, 1973-75.

Media   Etching, aquatint, mezzotint, painting.

One-person Exhibitions   Numerous, including: Associated American Artists, New York; Lenore Gray Gallery, Providence, Rhode Island, 1975; Edward Hopper Foundation, Nyack, New York, 1976; Brooklyn Collector, 1977.

Group Exhibitions   Yale University Art Gallery, New Haven; Brooklyn Museum; Boston Printmakers National Print Show; Cornell University; Society of American Graphic Artists, New York (4 years); Library of Congress, Washington, D.C.; Silvermine Guild of Artists, Inc., New Canaan, Connecticut; Princeton University Art Gallery; Smithsonian Institution traveling exhibition; Museum of Fine Arts, Cincinnati; Albany Art Gallery; Associated American Artists, New York.

Honors/Awards   Louis Comfort Tiffany Award, 1964; Yale University Norfolk Fellowship, 1963; Yale University Beinecke Grant, 1964; Purchase Prizes, Library of Congress, 1966, 1968, 1971; Purchase Prize, Yale University Art Gallery, 1966; Fulbright Grant, 1967; Research Foundation Grant, State of New York, 1968, 1969, 1973; Creative Artists Public Service Fellow, 1975.

Collections   Library of Congress; Yale University Art Gallery; Albright-Knox Museum, Buffalo, New York.

Commissions/Works Published in   *Artist's proof.* Prints for Associated American Artists, New York. Print editions.

30 "Daisies," *Rudbeckia* sp.
Etching, aquatint, and mezzotint;
23⅜ x 35⅝" plate mark

BEYER, Kurt

Born   Bromberg, Germany,
9 January 1896.

Died   Bad Liebenzell, 31 October
1973.

Education   Art and Craft Museum,
Berlin: teaching degree, 1925.

Career   Crafts teacher, illustrator,
painter.

Media   Watercolor, ink.

One-person Exhibitions   Oldenburg,
1951; Frankfurt, 1952; Museum at Müllheim, 1949, 1953,
1954, 1955, 1960; High School, Müllheim, 1952; Museum at
Essen, 1953, 1955; Rathaus, Hannover, 1959; Kubus,
Hannover, 1966, 1971; Freiburg, 1968, 1971, 1972.

Group Exhibitions   Orangerie, Hannover, 1966;
Galeriegebäude, Hannover, 1968.

Works Published in   Winkel, Gerhard. *Naturkunde für
Stadtschulen.* Frankfurt am Main, Verlag Moritz
Dietsterweg, n.d.
Boros, G. *Unsere Heil- und Teepflanzen.* Stuttgart, Eugen
Ulmer, 1963.
Boros, G. *Unsere Heil- und Teepflanzen II.* Stuttgart,
Eugen Ulmer, 1965.*
Boros, G. *Unsere Küchen- und Gewürzkräuter.* Stuttgart,
Eugen Ulmer, 1960, 1966.

31 Cranesbill, *Geranium sanguineum*
Watercolor; 6⅞ x 6⅝"

32 Ivy, *Hedera helix;* Black Alder, *Alnus glutinosa;*
Mouse-ear Hawkweed, *Hieracium pilosella* *
Ink; 10⅛ x 7"
Gift of Frau Gudrun Beyer

33 Betony, *Stachys officinalis;* Snakeweed,
Polygonum bistorta; Water Mint,
Mentha aquatica *
Ink; 10⅛ x 7"
Gift of Frau Gudrun Beyer

34 (Clockwise from top) *Silene rupestris;
Sempervivium montanum; S. alpinum;
Minuartia verna; Sempervivum* sp.;
Plantago alpina; Pedicularis kerneri
Watercolor; 11¾ x 5⅞"

34

BLOS, Maybelle (Mrs. Peter)

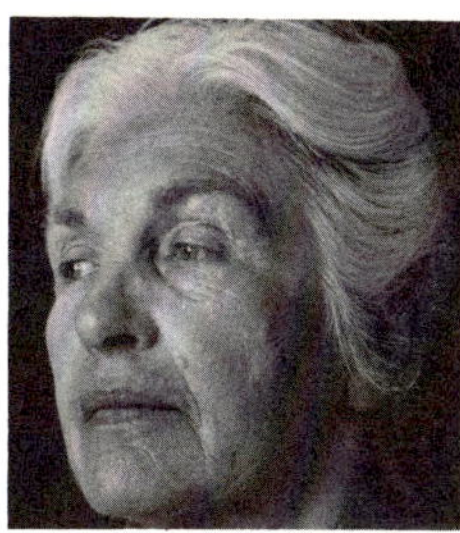

Born   Sebastopol, California,
1 May 1906.

Address   29 Live Oak Road,
Berkeley, California 94705.

Education   Studied botany with
Milo Baker, Santa Rosa Junior
College, Santa Rosa, California:
1924. University of California,
Berkeley: A.B., Art, 1926. Studied
art in Europe, 1927-31.

Career   Free-lance scientific
illustrator. Formerly, exhibit designer and illustrator,
National Parks Service, Museum Division, Berkeley, 1937-39.
Zoological illustrator, Museum of Vertebrate Zoology,
University of California, 1940-43. Part-time in Department
of Zoology, Paleontology, and the Botanic Garden, University
of California, Berkeley, 1940-72.

Media   Ink, acrylic.

Group Exhibitions   Hunt Institute International, 1964;
Canessa Gallery, San Francisco; Lawrence Hall of Science,
University of California.

Honors   Phi Beta Kappa, 1927; Taussing Fellowship,
University of California, Berkeley, 1927-28.

Collections   Huntington Library, Art Gallery, and Botanical
Gardens, San Marino, California; Regional Park Botanic
Garden with the Charles Lee Tilden Regional Park.

Commissions/Works Published in   Goodspeed, T.H. *The
genus Nicotiana*. Waltham, Massachusetts, Chronica
Botanica Press, 1954.
Cactus and succulent journal, 1952 through the 1960's.
Hall, E.R. *Mammals of Nevada*.
Hall, C. Unpublished book on ferns.
Murals, Paleo Museum, University of California.

35 *Corryocactus squarrosias*
Ink; 15 x 10⅛″

36 *Lobivia westii*
Ink; 14⅞ x 10″

37 *Borzicactus madisoniorum*
Ink; 14⅞ x 10½″

38 *Opuntia dimorpha*
Ink; 13 x 10⅛″

Gifts of Mr. Paul Hutchison

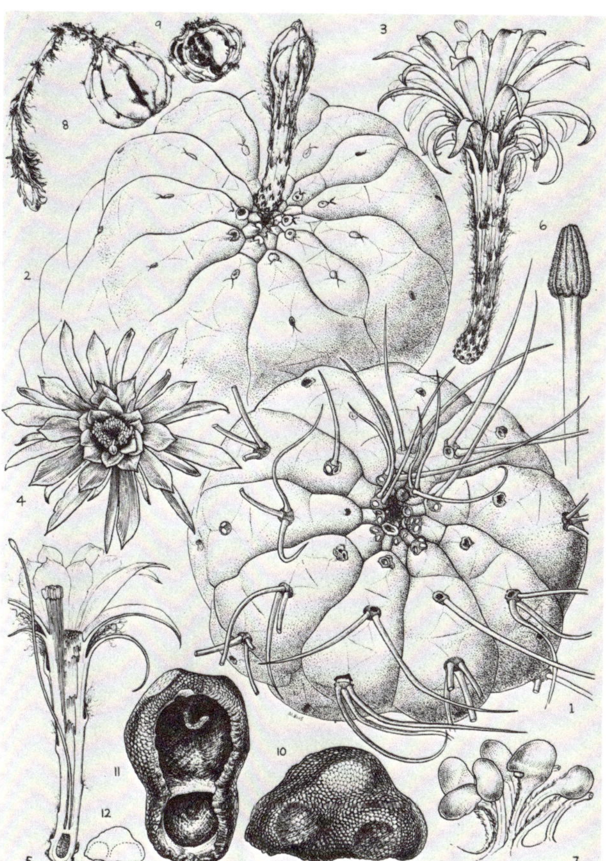

37

BOOTH, Raymond C.

Born   Leeds, Yorkshire, England,
8 August 1929.

Address   22 Far Moss, Leeds 17,
Yorkshire, England.

Education   Leeds College of Art:
National Diploma in Design,
1951; Art Teacher's Diploma,
1952.

Career   Free-lance artist.

Media   Oil, watercolor, pencil, ink,
acrylic, crayon, pastel.

One-person Exhibitions   Walker's Galleries, London, 1959;
The Fine Art Society, London, 1975.

Group Exhibitions   London: Royal Horticultural Society;
Fine Art Society Ltd.; Royal Academy Summer Show;
Society of Wildlife Artists. Hunt Institute International,
1972; Hunt Institute Travel Shows (Plants in Art,
International).

Awards   Bronze Grenfell Medal, Royal Horticultural Society,
1952; Grenfell Medal, Royal Horticultural Society, 1953.

Collections   Royal Horticultural Society; University of
Aberdeen.

Works Published in   Urquhart, B.L. (ed.). *The Camellia.*
London, Urquhart Press, Vol. 1, 1956; Vol. 2, 1960.
Synge, P.M. (ed.) *Some good garden plants.* London,
Royal Horticultural Society, 1962.
Illustrated London news (Christmas editions); *Journal of
the Royal Horticultural Society;* Royal Horticultural
Society yearbooks.

39 Ground Orchid, *Dactylorhiza elata*
Oil on paper; 26¾ x 12½″

40 Bellflower, *Campanula latifolia*
Oil on paper; 27⅛ x 16⅜″

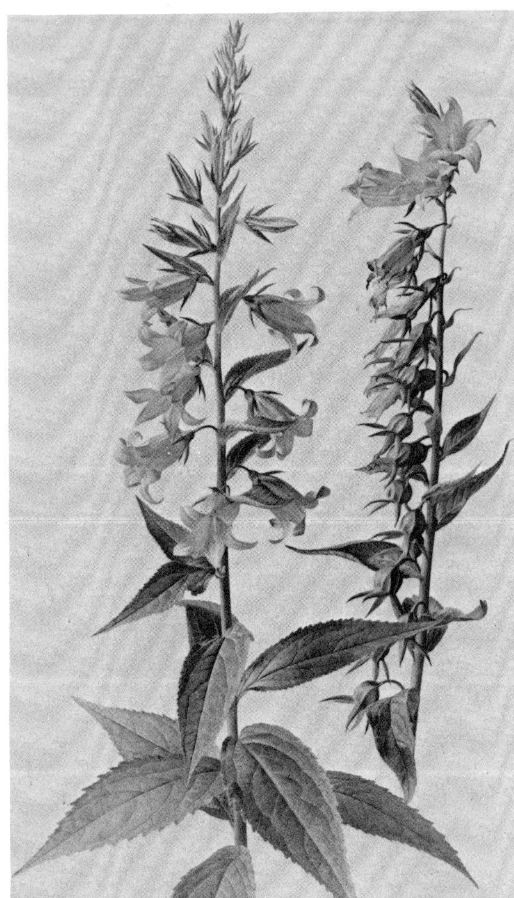

40

BORKOWSKY-BRAENDLIN, Hanni (Mrs. Hans)

Born   Wädenswil, Switzerland, 14 May 1902.

Address   Zürcherstrasse 145, 8640 Rapperswil St. Gallen, Switzerland.

Education   School of the Profession of Art, Zürich: degree in textile design, 1921. Academy of Arts, Karlsruhe, Germany: 1922-23.

Medium   Watercolor.

One-person Exhibitions   Orell-Füssli Gallery, Zürich, 1971; Heimatmuseum, Rorschach, 1973; Schloss, Rapperswil, 1975.

Group Exhibitions   Hunt Institute International, 1972; Hunt Institute Travel Shows (International).

Works Published in   *Du,* 1969; *St. Galler Volksblatt,* 1969, 1971; *Die Linth,* 1969, 1971, 1975; *Die Tat Volksblatt,* 1971; *Neue Zürichzeitung,* 1971; *Ostschweizer Tagblatt,* 1973; *Blumenmappe, Aus meinem Garten,* 1975; *Zürichseezeitung,* 1975.

41 "Mosses and lichens from the Engadin"
Watercolor; 9⅝ x 8¼"

42 Auricula, *Primula auricula* hybrid
Watercolor; 11½ x 8⅞"
Both lent by the artist

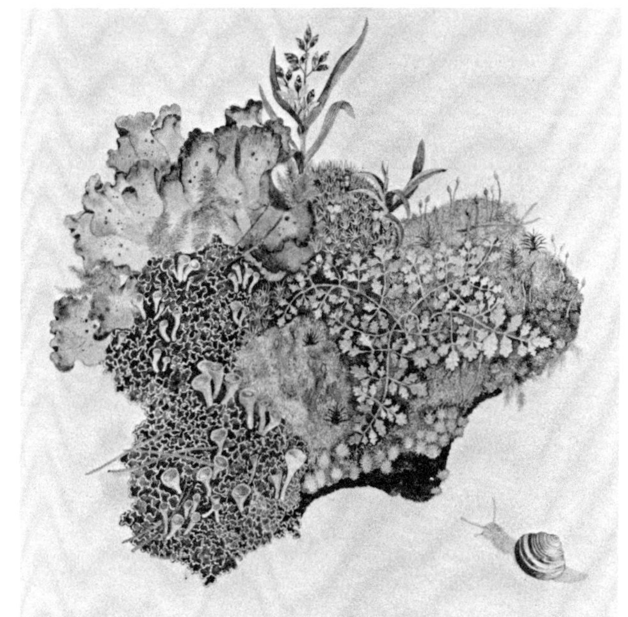

41

BUTLER, Mary

Born   Denver, Colorado, 28 June 1919.

Address   6610 West Olympic Boulevard, Los Angeles, California 90048.

Education   Kirkland School, Denver. Chouinard Art Institute, Los Angeles.

Career   Staff artist, Los Angeles County Museum of Natural History, 1961 to present. Formerly, owner of commercial art service, Denver. Free-lance illustrator for publishing companies in Los Angeles and New York, 1948-61.

Media   Watercolor, oil, acrylic, pastel, ink, scratchboard, airbrush.

Commissions/Works Published in   Science textbooks published by L.W. Singer, Syracuse, New York.
La Brea story. Los Angeles, Ward Richie Press.
Fossil vertebrates of southern California. Berkeley, University of California Press.
Bizarre plants. New York, Macmillan Publishing Company.
Numerous illustrations for scientific papers.
Paintings of Pleistocene animals for Hancock Park, Los Angels and also used in the film, *Permafrost Frontier*, produced by the Aleyska Pipeline Company.

43 Orchid, *Catasetum fimbriatum*
    Gouache; 10¾ x 8½″

44 Screw-Pine, *Pandanus* sp.
    Gouache; 9½ x 7″

44

CAWEIN, Kathrin

Born   New London, Connecticut, 9 May 1895.

Address   35 Mountain Road, Pleasantville, New York 10570.

Education   Art Students League, New York: 1927-32. Oberlin College, Ohio: M.A.

Career   Printmaker and calligrapher. Formerly, *Music Roll* editor, 1910-35. Graphics instructor, County Center, White Plains, New York, 1935-36. Children's art instructor in home studio, 1950-53.

Media   Graphic arts, pastel, calligraphy.

One-person Exhibitions   County Center, 1935; Town Hall, New York, 1950; 8th St. Playhouse, New York, 1953; University of Tampa, Florida, 1973; Sarasota, Florida, 1973; Oberlin College, 1974; St. John's Church, Pleasantville, 1976; Berea College, Kentucky, 1977.

Group Exhibitions   Century of Progress, 1934; Texas Centennial, 1937; New York World's Fair, 1939; and in England, France, Italy, and Ecuador.

Awards   Frank Talcot Prize, Society of American Etchers, 1936; Lithograph Prize, Village Art Center, 1944; Etching Prize, National Association of Women Artists, 1947. Awards for drypoint: Pleasantville Women's Club, 1950; Hudson Valley Art Center, 1951; Westchester Federation Women's Club, 1952.

Collections   National Museum, Washington, D.C.; Metropolitan Museum of Art, New York; Oberlin College; University of Tampa. Manuscripts in: St. Mark's Church, Van Nuys, California; St. Andrew's Church, Sarasota, Florida; and the Congregational Church, Colorado Springs.

45 "The Veteran of Pilot Knob"
Lithograph; 15 1/16 x 10 1/4"

CHEN, Chien-chu

Born   Taipei, Taiwan, 18 February 1927.

Address   Department of Botany, School of Sciences, National Taiwan University, Taipei 107, Taiwan.

Career   Herbarium Assistant, Department of Botany, National Taiwan University. Specialist in plant anatomy and scientific drawing. Formerly, free-lance botanical illustrator, 1956-65.

Media   Ink, gouache, watercolor.

Group Exhibitions   Hunt Institute International, 1972; Hunt Institute Travel Show (International); Hunterdon Art Center, Clinton, New Jersey, 1977.

Honor   Member of Artists Association of The Republic of China.

Works Published in   Hsieh, A.-tsai and Chen, Chien-chu. *Medicinal trees*. Taipei, Taiwan Chung Hwa Forestry Association, Vol. I, 1956; Vol. II, 1957.

Lin, Wei-chih. *Study on the classification of Bambusaceae in Taiwan*. Taipei, Taiwan Forestry Research Institute, 1961.

Lui, T.S. *Illustrations of native and introduced ligneous plants of Taiwan*. Taipei, National Taiwan University, 1962.

Li, H.L. *Woody flora of Taiwan*. Narberth, Pennsylvania, Livingston Publishing Co., 1963.

Lin, Wei-chih. *New bamboos from Taiwan*. Taipei, Taiwan Forestry Research Institute, 1964.

Chuan, Ching-chang and Huang, Chia. *The Leguminosae of Taiwan and Taiwania*. Taipei, National Taiwan University, 1965.

Lin, Wei-chih. "Studies on morphology of bamboo flowers," *Taiwan Forestry Research Institute bulletin*, No. 248, 1974.

Lin, Wei-chih. "The classification of subfamily Bambusoideae in Taiwan," *Taiwan Forestry Research Institute bulletin*, No. 271, 1976.

Flora of Taiwan. Taipei, Epoch Publishing Co., Vol. I-III, 1975-77.

46 *Melastoma candidum*
Gouache; 10½ x 8″

47 *Typhonium divaricatum*
Gouache; 9½ x 8¼″

Both lent by the artist

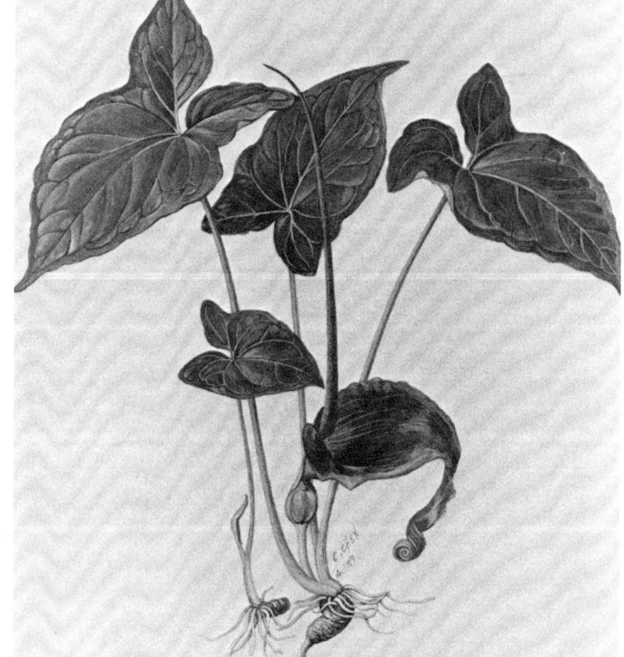

47

CLEUTER, Albert

Born   Brussels, Belgium, 10 August 1901.

Address   56 rue Alexandre Markelbach, 1030 Brussels, Belgium.

Education   Académie Royal des Beaux Arts, Brussels: Architect Diploma, 1922.

Career   Botanical illustrator, Jardin Botanique National, Brussels, 1936 to present.

Media   Ink, watercolor, color pencil.

Group Exhibitions   100th Anniversary Exhibition, Jardin Botanique National, 1970; Ars Botanica, Brussels, 1968; Hunt Institute Internationals, 1968, 1972.

Works Published in   Fouarge, J. and Louis, J. *Les grandes essences forestières d'Afrique.* Brussels, Institut National pour l'Etude Agronomique du Congo, 1947.
Robyns, W. (ed.). *Flore du Congo Belge et du Ruanda-Urundi.* Brussels, Jardin Botanique de l'État, in volumes since 1948.
Lawalrée, A. *Flore generale de Belgique.* Brussels, Ministère de l'Agriculture, 1950-59.
Espaces verts et arts des jardins. Brussels.
Bulletin du Jardin Botanique National de Belgique.
Flore générale de Belgique.
Dioramas of Ruwenzori, Uganda.

48 Dropwort, *Filipendula hexapetala*
Ink; 15½ x 9⅝″

49 *Cotoneaster integerrima*
Ink; 15 x 9⅛″

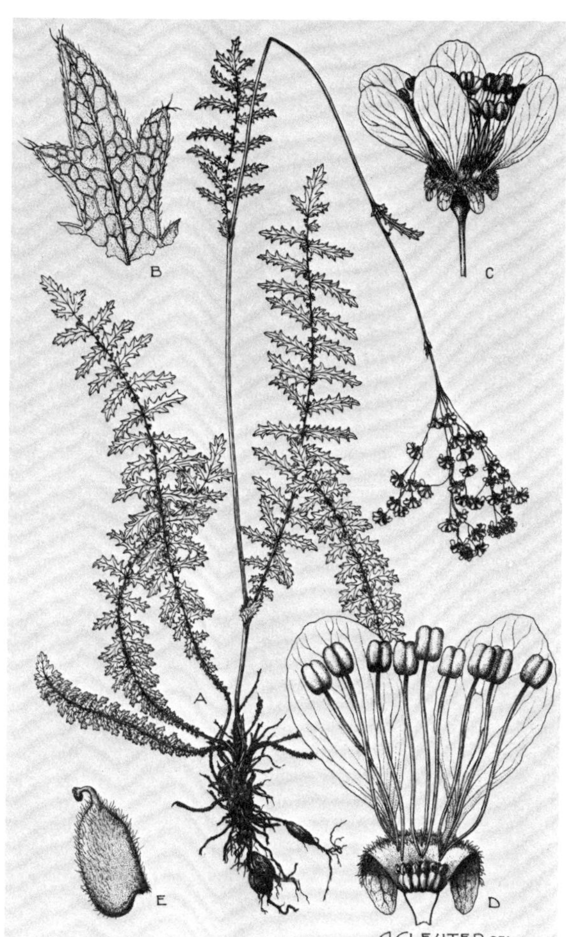

48

CLEVELAND, Walter

Born Santa Barbara, California, 29 February 1940.

Address 2448 Westlane Avenue North, Seattle, Washington 98109.

Education Pennsylvania Academy of the Fine Arts, Philadelphia; studied printmaking, 1961. Pasadena City College: 1962-69.

Career Self-employed artist.

Media Oil, watercolor, drawing, etching.

Group Exhibitions Pennsylvania Academy of the Fine Arts, 1960-61; Los Angeles County Art Museum, 1962; California State Fair (2 years); Small Images II, California State University, Los Angeles, 1967; Ithaca College of Art Museum, New York; United States Information Agency, overseas embassies; Hunt Institute International, 1972; Hunt Institute Travel Shows (International, Botanical Prints); 5 Westcoast Printmakers, Hunt Institute, 1975-76.

Collections United States Information Agency; many private collections.

Works Published in Print editions.

50 Foxglove, *Digitalis* sp.
   Color etching; 10 x 14″ plate mark

51 Morning Glory and Wild Grass
   Color etching; 10 x 15″ plate mark

50

CONNEEN, Jane W. (Mrs. Joseph L.)

Born   Montclair, New Jersey, 19 April 1921.

Address   The Little Farm Workshop, Box 279, R.D. 1, Bath, Pennsylvania 18014.

Education   The Masters School, Dobbs Ferry, New York: 1934-37. Studied oil painting with George Parker, New York, 1937-40. Studied oil painting with Emily Hatch, New York, 1940-41. Completed Famous Artists' Commercial Course, 1961.

Career   Free-lance artist and illustrator.

Media   Woodcut, linocut, etching, ink, watercolor.

One-person Exhibitions   Rogues Gallery, Allentown, Pennsylvania, 1973; Dubois Gallery, Lehigh University, Bethlehem, Pennsylvania, 1975; Wildflowers Gallery, Philadelphia, 1975; Kemerer Museum, Bethlehem, 1976.

Group Exhibitions   Allentown Art Museum, 1971-76; Lehigh Valley Art Alliance, 1972-77; Philadelphia Civic Center, 1973; Hunterdon Art Center, Clinton, New Jersey, 1974-77; Old York Road Art Guild, 1974; Muhlenberg College, Allentown, 1974; Doylestown Art League, Doylestown, Pennsylvania, 1975; Phillip's Mill Art Exhibition, New Hope, Pennsylvania, 1975; Miniature Painters, Sculptors and Gravers Society of Washington, D.C., 1975, 1976; Miniature Art Society of Florida, 1976, 1977; Miniature Art Society of New Jersey, 1976, 1977.

Awards   Second Prize, Graphics, 1972, and Honorable Mention, 1975-77, Lehigh Valley Art Alliance; Second Prize, Watercolor, Art in an Arboretum, Allentown, 1976; First Prize, Printmaking, Bicentennial Art Show, Mansfield Township, New Jersey, 1976.

Collections   A.T. & T. Long Lines, Bedminster, New Jersey; numerous private collections.

Commissions   Annual brochure, Allentown Art Museum, 1972. Christmas card designs for Abel Cards, Media, Pennsylvania, 1973.
Miniature book on wildflowers. Winterport, Maine, The Borrower's Press [in progress].

52 Rieger Begonia
Ink and watercolor; 14 x 10″

53 Red Geranium, *Pelargonium* hybrid
Ink and watercolor; 12¾ x 10″

54 Geranium, *Pelargonium* hybrid
Ink; 13½ x 10″

53

CRAWFORD, Kathleen S.

Born   Oakland, California, 16 July 1910.

Address   1153 16th Street, Los Osos, California 93402.

Education   Chouinard Institute of the Arts, Los Angeles: 1927. Self-taught artist.

Career   Free-lance artist.

Media   Casein watercolor, tempera.

One-person Exhibitions   Academy of Sciences, San Francisco; San Diego Natural History Museum; Los Angeles Natural History Museum, 1972; Missouri Botanic Garden; Brooklyn Botanic Garden; Biomedical Library, University of California, Los Angeles, 1973; Santa Barbara Natural History Museum, 1974.

Collection   Santa Barbara Natural History Museum.

Commissions/Works Published in   *Nature,* March 1955.
Travel, March 1971.
Santa Barbara news press.
Pasadena star news.
Christmas card designs for Chryson's.

55 Joshua Tree, *Yucca brevifolia*
Casein watercolor; 24¼ x 14¾″

56 Pond Lily, *Nuphar advenum*
Casein watercolor; 29 x 19½″

56

DEAN, Ethel

Born   New York, New York,
5 September 1908.

Died   30 May 1971.

Education   Parson's School of
Design, New York: 1925-26.

Career   Designer of fabrics,
wallpapers, interiors, theater
costumes, and architecture.

Media   Gouache, watercolor, ink,
pencil.

Award   For illustrations in the *Rubaiyat*.

Collections   Cooper-Hewitt Museum of Design, New York;
Smithsonian Institution, Washington, D.C.; Central Library
for Music and Dance, Tel-Aviv.

Commissions/Works Published in   Fabrics and wallpapers.

57 "Woodland Wildflowers,"
Fabric design
Gouache; 26½ x 19¼"

58 Fabric design
Gouache; 27 x 21"

Gifts of Mr. Abner Dean

57

DE GEUS, Marie

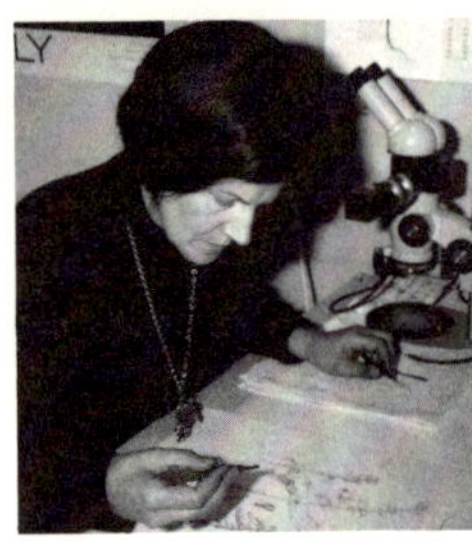

Born   Heerhugowaard,
Netherlands, 20 July 1934.

Address   Achterdorpsstraat 55,
Renkum, Netherlands.

Education   Studied with M. Pronk
Rompelman, Haarlem, and with
Vok Koster.

Career   Botanical artist, State
Agricultural University,
Wageningen, 1970 to present.

Works Published in   Botanical
periodicals.

59 *Cicer nuristanicum*
Ink; 8 x 5½″

60 *Dichapetalum arachnoideum*
Ink; 13⁷⁄₁₆ x 9⅛″

Permanent loans of the Laboratorium voor
Plantensystematiek en -Geografie,
Landbouwhogeschool, Wageningen, Netherlands

60

DERMEK, Aurel

Born   Brodské, Czechoslovakia, 6 July 1925.

Address   Bullova 3, 83000 Bratislava, Czechoslovakia.

Education   Technical Building High School, Bratislava: 1941-46.

Career   Building designer, free-lance botanical artist, mycologist, author.

Medium   Watercolor.

One-person Exhibition   Slovak National Museum, Bratislava, 1973.

Collection   Slovak National Museum, Bratislava.

Works Published in   Dermek, A. *Naše huby (Our mushrooms)*. Bratislava, Obzor, 1967.
Dermek, A. "Naše kozáky" ("Our Leccinums"), *Živa*, Vol. 17, 1969.
Dermek, A. and Pilát, A. *Poznávajme huby (How to know the mushrooms)*. Bratislava, Veda, 1974. *
Pilát, A. and Dermek, A. *Hribovité huby (The boletes)*. Bratislava, Veda, 1974.
Dermek, A. *Huby lesov, polí a lúk (Mushrooms of the woods, fields and meadows)*. Martin, Osveta, 1976.
Dermek, A. *Colored illustrations of larger fungi of Slovakia* [in progress].

61 Fleshy Pore Mushroom, *Boletus calopus**
Watercolor; 10 x 7½″

62 Fleshy Pore Mushroom, *Leccinum percandidum**
Watercolor; 9⅞ x 7¼″

63 Fleshy Pore Mushroom, *Boletus erythropus**
Watercolor; 10 x 7⅜″

64 Fleshy Pore Mushroom, *Aureoboletus gentilis**
Watercolor; 9¾ x 7⅜″

65 Fleshy Pore Mushrooms, *Boletus rhodoxanthus**
and *B. rhodopureus*
Watercolor; 10⅛ x 7⅜″

66 Fleshy Pore Mushroom, *Leccinum griseum**
Watercolor; 10⅛ x 7¼″

66

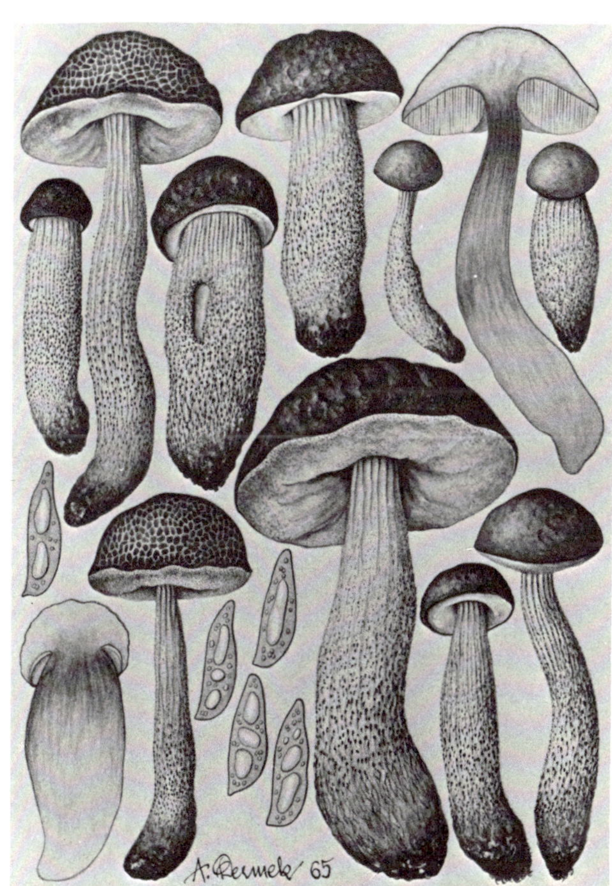

"

DICKSON, Lalia C. P.–

Born   Edinburgh, Scotland,
   26 January 1913.

Address   1, Inverleith Row,
   Edinburgh, EH3 5LP, Scotland.

Education   Edinburgh College of
   Art: Diploma Course, 1930-34.

Career   Free-lance illustrator.

Media   Watercolor, pencil, oil,
   oil pastel, charcoal, gouache.

One-person Exhibitions   Gladstone's
Land, Edinburgh, 1963; Douglas and Foulis Gallery,
Edinburgh, 1970, 1973, 1974; Gallery Paton, Edinburgh, 1974.

Group Exhibitions   Numerous.

Collections   South East Scotland Regional Hospital Board;
many private collections.

Commissions/Works Published in   *Turkish Flora.* Vol III,
   Edinburgh Department of Botany, Edinburgh University.
   Wild Flowers. Books I-IV, W. & R. Chambers, Ltd.
   Numerous commissions, including paintings for
   The Medici Society, Ltd.

67   Primrose, *Primula denticulata*
   Watercolor; 16 x 11¾"

68   "Midsummer Study in the Rockery,
   Edinburgh, Scotland"
   Watercolor; 14 x 9¼"
   Lent by the artist

68

DOUTHIT, Karin M. (Mrs. Harry)

Born   Essen, Germany, 2 November 1930. Has worked in the United States since 1961.

Address   801 Fifth Street, Ann Arbor, Michigan 48103.

Education   Textilingenieurschule, Krefeld, Germany: 1947-49. Landeskunstschule, Hamburg, 1949-51.

Career   Botanical illustrator, University of Michigan, Ann Arbor, 1969 to present.

Media   Pencil, watercolor, ink, acrylic.

One-person Exhibition   Del Mar College, Corpus Christi, Texas, 1968.

Group Exhibitions   Del Mar College, Corpus Christi, 1967-68; Art Guild, Madison, Wisconsin, 1967; Hunt Institute International, 1972; Hunt Institute Travel Shows (International, Decorative Arrangements); Hunterdon Art Center, Clinton, New Jersey, 1977.

Awards   1st Award for Drawing, Del Mar College, 1967; 1st Award for Drawing, Art Guild Exhibition, 1967; 4th Award for Drawing, Del Mar College, 1968; 2nd Award, Hunterdon Art Center, 1977.

Works Published in   *Brittonia*, Vol. 14, 1962.
Rhodora, No. 66, 1964.
Contributions from the University of Michigan Herbarium, Vol. 9, 1972; * Vol. 11, 1974; Vol. 12, 1975.
American journal of botany, No. 59, 1972.

69 *Montanoa laskowskii* *
   Ink; 13⅜ x 9⅞″

70 *Otopappus jaliscensis*
   Ink; 13⅝ x 9¾″

71 *Verbesina mickelii* *
   Ink; 13⅞ x 9¾″

72 Groundsel, *Senecio galicianus* *
   Ink; 13¾ x 9⅛″

73 "Bouquet II"
   Pencil; 21½ x 14¾″
   Lent by the artist

69

DOWDEN, Anne Ophelia Todd (Mrs. Raymond B.)

Born   Denver, Colorado,
17 September 1907.

Address   205 West 15th Street,
New York, New York 10011.

Education   University of Colorado,
1925-26. Carnegie Institute of
Technology, Pittsburgh: B.A.,
1930. Beaux Arts Institute of
Design. Art Students League,
New York.

Career   Free-lance botanical artist
and writer, 1950 to present. Formerly, drawing instructor,
Pratt Institute, New York, 1930-32. Chairman, Art
Department, Manhattanville College, New York, 1932-53.
Free-lance textile designer, New York, 1935-55.

Medium   Watercolor.

One-person Exhibitions   Silvermine Guild of Artists, Inc.,
New Canaan, Connecticut, 1955; Brooklyn Botanic Garden,
1957, 1977; Hunt Institute, 1965; Hunt Institute Travel
Shows (Shakespeare's Flowers).

Group Exhibitions   Numerous, including: Metropolitan
Museum of Art, New York; Carnegie Institute, Pittsburgh;
Hunt Institute Internationals, 1964, 1968, 1972; Hunt
Institute Travel Shows (International).

Honors/Awards   Tiffany Foundation fellowships, 1929, 1930,
1932; Order of the Delta Gamma Rose, 1967; numerous
awards for books.

Collections   Brooklyn Botanic Garden; New York Botanical
Garden; Callaway Gardens, Georgia; Holden Arboretum,
Ohio.

Commissions/Works Published in   Dowden, A.O. *Cuas 8, the
little hill.* Cooper Union, 1961.
Dowden, A.O. *Look at a flower.* New York, Thomas Y.
Crowell, 1963.
Dowden, A.O. *The secret life of the flowers.* New York,
Odyssey Press, 1964.

Dowden, A.O. and Thomson, R. *Roses.* New York, Odyssey
Press, 1965.
Wildflowers of the United States. New York Botanical
Garden and McGraw-Hill, all volumes, 1966 to present.
Borland, H. and Dowden, A.O. *Plants of Christmas.* New
York, Golden Press, 1969.
Kerr, J. *Shakespeare's flowers.* New York, Thomas Y.
Crowell, 1969.
Untermeyer, L. *Plants of the Bible.* New York, Golden
Press, 1970.
Dowden, A.O. *Wild green things in the city: a book of
weeds.* New York, Thomas Y. Crowell, 1975. *
Dowden, A.O. *The blossom on the bough: a book of trees.*
New York, Thomas Y. Crowell, 1975. **
Busch, P. *Wildflowers and the stories behind their names.*
New York, Charles Scribner's Sons, 1977.
Borland, H. *The golden circle: a book of months.* New
York, Thomas Y. Crowell, 1977.
Collector print series. Louisville, Frame House Gallery,
1969-77.
*Audubon, House beautiful, Life, Natural history, The
garden journal* of New York Botanical Garden.
Paintings for Callaway Gardens, Holden Arboretum, and
New York Botanical Garden.

74 "Dried Seed Pods"*
Watercolor; 11 x 7⅞"

75 Sycamore, *Plantanus occidentalis***
Watercolor; 13½ x 12"

75

EVANS, Henry

Born   Superior, Wisconsin, 16 May 1918.

Address   555 Sutter Street, San Francisco, California 94102.

Education   San Francisco Junior College. San Francisco State College. University of California, Berkeley. University of Arizona, Tucson: B.A., 1942. Self-taught botanical artist.

Career   Botanical printmaker, 1958 to present. Formerly, bookseller in Tucson and San Francisco. Founded Peregrine Press which operated from 1949 to 1958.

Medium   Linocut.

One-person Exhibitions   Over 200 across the country and around the world, including: Smithsonian Institution; Royal Horticultural Society, London; California Academy of Sciences, San Francisco; Ha'aretz Museum, Tel Aviv; Field Museum, Chicago; Los Angeles County Museum; National Arboretum, Washington, D.C.; Hunt Institute Travel Shows (State Flowers), Hunt Institute, 1966.

Group Exhibitions   Numerous, including: Garden Center of Greater Cleveland; XIth Botanical Congress, Seattle; Hunt Institute Internationals, 1964, 1968, 1972; 5 Westcoast Printmakers, Hunt Institute, 1976; Hunt Institute Travel Shows (International, Plants in Art, Roses, Lilies, Botanical Prints).

Collections   Numerous, including: San Francisco Public Library; Morton Arboretum, Lisle, Illinois; New York Botanical Garden; Missouri Botanical Garden; Garden Center of Greater Cleveland; National Arboretum, Washington, D.C.; Northwestern University, Chicago; Dartmouth College, Hanover, New Hampshire; Library of Congress, Washington, D.C.; many private collections.

Commissions/Works Published in   Strabo, W. *Hortulus.* Pittsburgh, Hunt Institute, 1966.
Bulla, C.R. *Flowerpot gardens.* New York, Thomas Y. Crowell Co., 1967.
Evans, H. *State flowers.* San Francisco, Henry Evans, 5 Vols., 1968-72.
Kooy, J. "Henry Evans, printmaker," *Pacific discovery,* 1970, 1974.
Evans, H. *Botanical prints with excerpts from the author's notebooks.* San Francisco, W.H. Freeman and Co., 1977.
Evans, H. *Prints.* San Francisco, Henry Evans, 37 Vols., 1959-77. *
Pacific discovery, San Francisco chronicle, Cleveland explorer, California Horticultural Society quarterly.
Triptych for the 927 Gallery, New Orleans, 1977.

76 "Violet," *Viola* sp*
   Linocut; 10½ x 16½"

77 "Crocus," *Crocus* sp.*
   Linocut; 10¼ x 16"

Editions printed by Marsha Evans

76

FAWCETT, Priscilla

Born   Blewbury, Berkshire, England, 20 June 1932.

Address   Fairchild Tropical Garden, 10901 Old Cutler Road, Miami, Florida 33156.

Education   St. Mary's School, Cuckfield Park, Sussex: domestic science, 1948-49. British Museum of Natural History: studied scientific illustration under the direction of Dr. William T. Stearn, 1963.

Career   Botanical artist, Fairchild Tropical Garden, 1965 to present. Formerly, free-lance artist.

Media   Ink, watercolor.

One-person Exhibitions   Neiman-Marcus, Miami Beach, 1972; Miami Art Center, 1974; Callaway Gardens, Georgia, 1977.

Group Exhibitions   Hunt Institute International, 1968; International Exhibition of Botanical Art, Johannesburg, 1973.

Works Published in   Dodson, C.H. and Gillespie, R.J. *The biology of the orchids*. Nashville, Mid-America Orchid Conference, 1967.
Jamieson, B.G.M. and Reynolds, J.F. *Tropical plant types*. London, Pergamon Press, 1967.
van der Pijl, L. and Dodson, C.H. *Orchids, flowers, their pollination and evolution*. Miami, University of Miami Press, 1968.
Norstog, K. and Long, R. *Plant biology*. Philadelphia, W.B. Saunders Company, 1976.
Tomlinson, G. *South Florida trees* [in progress].
Correll, D. *Flora of the Bahamas* [in progress].

78 Passion Flower, *Passiflora coccinea*
Watercolor; 27 x 16″

79 Passion Flower, *Passiflora edulis* var. *flavicarpa*
Watercolor; 27⅝ x 17⅝″
Lent by the artist

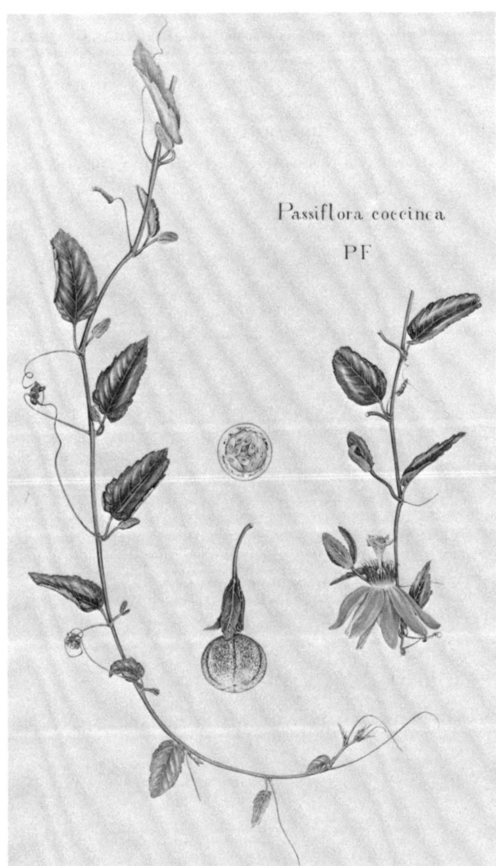

78

FELSKO-SCHÜLKE, Elsa

Born   Berlin, Germany,
2 September 1908.

Address   Theissbaum—Weg 4,
5108 Monschau-Kalterherberg,
Germany.

Education   Botanisches Museum,
Berlin-Dahlem: botanical
supervision by Professors Dr.
Robert·Pilger and Dr. Hermann
Reimers, 1950-54.

Career   Free-lance artist and
illustrator, 1954 to present. Formerly, embroidery designer
for a textile firm, prior to 1942. Graphic artist, Deutscher
Normenausschuss (German Bureau of Standards), 1942-45.
Botanical artist, Botanisches Museum, Berlin-Dahlem,
1950-54.

Media   Watercolor, ink, oil.

One-person Exhibitions   Universitätsbuchhandlung Dietrich,
Göttingen, 1950; Botanisches Schaumuseum, Berlin, 1975;
Hirsch-Apotheke, Aachen, 1975.

Group Exhibitions   Deutsche Gartenbaugesellschaft, Berlin,
1950; Botanisches Schaumuseum, Berlin, 1963; Tryon Gallery,
London, 1968; Hunt Institute Internationals, 1968, 1972;
Galerie Taube, Berlin, 1973-74; International Exhibition of
Botanical Art, Johannesburg, 1973; Hunterdon Art Center,
Clinton, New Jersey, 1977.

Collections   Botanisches Museum, Berlin-Dahlem; University
of Göttingen; Herbig-Verlag, Munich; Deutsche Schell AG.,
Hamburg.

Works Published in   Reimers, H. *Blumen-Atlas.* Berlin, F.A.
Herbig-Verlag, 1950.
Herbig-Kalender. Berlin, 1954-74.

Irion. *Drogisten-Lexikon.* Berlin, Springer-Verlag, 1955.
Littleboy, S. and Clokie, H. *A book of wild flowers.* Oxford,
Bruno Cassirer, 1956 (reprinted in 1963).
Muscheln gaben die Namen. Berlin, Deutsche Schell A.G.,
1956.
Littleboy, S. and Clokie H. *Portraits of wild flowers.*
Oxford, Bruno Cassirer, 1959.
Universe calendar of flowers. New York, Universe Books,
Inc., 1960.
Gäbler, H. *Das Büchlein von den Heilenden Kräutern.*
Berlin, F.A. Herbig-Verlag, 1963, and Munich, Goldmann
Verlag, 1970.
Reimers, H. *Blumenfibel.* Berlin, F.A. Herbig-Verlag, 1967
(revised edition, 1976).
Portfolios: *Blumenbilder,* 1970; *Müttergenesungswerk-Karten,*
1970.
Questri nostri fiori. Bologna, Casa Editrice Capitol, 1970.
Schmeil Pflanzenkunde. Westermann's Monatshefte.

80 "Alter Baum in Garten Gesthemane,"
Olive, *Olea europaea*
Ink; 16¾ x 13⅛"
Lent by the artist

81 French Rose, *Rosa gallica*
Watercolor; 12½ x 9"
Gift of the artist

80

FRAZIER, Vivien Lee

Born   Dallas, Texas.

Address   6466 Sudbury Lane, Dallas, Texas 75214.

Education   Dallas Art Institute.

Career   Free-lance botanical artist and illustrator. Formerly, Vice President, American Title Company, Dallas.

Media   Ink, watercolor, oil.

Group Exhibitions   Arthur A. Everts Art Show, 1966; Zonta Club of Dallas Art Show, Richardson, Texas, 1970.

Award   First Prize, Arthur A. Everts Art Show, 1966.

Collections   Texas Research Foundation, Renner, Texas; Environmental Protection Agency; many private collections.

Commissions/Works Published in   Correll, D.S. *The potato and its wild relatives.* Renner, Texas, Texas Research Foundation, 1962.
Lundell, C.L. *The genus Parathesis of the Myrsinacea.* Renner, Texas, Texas Research Foundation, 1966.
Correll, D.S. *Flora of Peru.* Chicago, Field Museum of Natural History, Vol. 13, 1967.
Correll, D.S. and H.B. *Aquatic and wetland plants of southwestern United States.* Washington, D.C., Environmental Protection Agency, 1972 (reissued by Stanford University Press, 1975).
Various issues of *Wrightii,* 1963-68.
Book jackets and illustrations for the Story Book Press, 1950-53.
Illustrations for Texas Research Foundation, 1958-71.
Many private commissions.

82 Hoary Azalea, *Rhododendron canescens*
Watercolor and ink on acetate;
14¹¹⁄₁₆ x 11¾″

83 Southern Wild Rice, *Zizaniopsis miliacza*
Watercolor and ink on acetate; 14¾ x 11¾″

Gifts of the artist

82

FUKUDA, Haruto

Born   Tokyo, Japan, 22 April 1936.

Address   56 Thomas Street, New York, New York 10013.

Education   Tokyo University of Fine Arts: M.A., 1963.

Career   Free-lance botanical illustrator.

Medium   Ink.

Collection   New York Botanical Garden.

Commissions/Works Published in   Flora Neotropica Organization (ed.). *Flora neotropica.* New York, Hafner Publishing Company, 1968.
Memoirs of the New York Botanical Garden, Vol. 23, 1972.
Koyama, T. "New Cyperaceae from Nepal and adjoining Tibet," *Botanical magazine* (Tokyo), 1973.
Koyama, T. "Smilacaceae," *Flora of Thailand,* 1975.*

84 *Heterosmilax polyandra**
   Ink; 11 x 7⅞″

85 Greenbrier, *Smilax magalantha**
   Ink; 10½ x 7½″

Both lent by Dr. Tetsuo M. Koyama

85

FUSI, Almina Dovati

Born   Carrara, Italy, 26 September 1908.

Address   Via Ninfale Fiesolano 14—50135, Florence, Italy.

Education   Academy of Fine Arts of Florence.

Media   Etching, copper engraving.

One-person Exhibitions   Carrara, 1932, 1943; La Spezia, 1937; Florence, 1958, 1961, 1965, 1971; Genoa, 1964; Rovereto, 1966; Bologna, 1967; Forte dei Marmi, 1973; Teglio in Valtellina, 1974.

Group Exhibitions   National Biennial Exhibitions of Engravers at Reggio, Emilia, Taranto, Padua, and Florence; National Exhibitions of Engravers, at Cremona, Rome, Cagliari, Soragna, Arezzo, Genoa, Milan, Turin, Ravenna, Forlì, and the Civic Museum of Pistoia; International Biennial Exhibitions of Graphic Arts, at Pescia and the Strozzi Palace, Florence; Third European Salon, at Nancy, Stockholm, and Mulhouse; International of Business and Professional Women, New York; Italian Contemporary, in principal cities of New Zealand; Acquisitions, 1944-74, Uffizi Gallery, Florence.

Awards   Gold Medals at exhibitions in Valombrosa, Naples, Florence, Osimo, Carrara, Palermo, and Marina di Pisa.

Collections   Uffizi Gallery, Florence; Historical Museum of the Navy, Rome; Museum of Engravings, Verona; Academy of Art, Montecatini; Franciscan Art Gallery, Assisi; Franciscan Gallery, Cortona.

Works Published in   Servolini, L. *The engravers of Italy.* 1960.
Roncaglia messenger, 1961, 1964.
Comanducci, 1962-72.
Bolaffi graphics, 1969-72.
Italian paintings, 1970.
Bolaffi art, 1973.
Numerous journals, annuals, and catalogues.

86   "Quasi Farfalle," Poppy, *Papaver* sp.
Etching; 18¼ x 7⅛" plate mark
Gift of the artist

GARDNER, Sue (Mrs. David)

Born   Woodward, Oklahoma, 26 May 1939.

Address   4637 Arlene Drive, Corpus Christi, Texas 78411.

Education   Studied printmaking at Del Mar College, Corpus Christi, and Texas A & I University, Kingsville, Texas. Del Mar College: B.S., Biology. Presently enrolled as a graduate student, Department of Horticultural Sciences, Texas A & M University, College Station, Texas.

Career   Botanist, botanical artist, and frequent contributor to the *Journal* of the Bromeliad Society, Inc. On Board of Directors, Bromeliad Society, Inc. Formerly, Editor, Corpus Christi Bromeliad Society *Bulletin*, 1970-76.

Media   Ink, acrylic, serigraphy, etching.

Group Exhibitions   World Bromeliad Show, Houston, 1972; Southwest Regional Orchid Growers Association Show, Corpus Christi, 1975; World Bromeliad Show, Los Angeles, 1975.

Commissions/Works Published in   Yang, L. *The terrace gardener's handbook.* New York, Doubleday & Company, Inc., 1975.
Marie Selby Botanical Gardens *Bulletin.*
Star to Star.
Note cards and a drawing on copper for the Bromeliad Society, Inc.*

87   *Tillandsia streptophylla**
Ink; 16¾ x 11½″

88   *Tillandsia flexuosa**
Ink; 20½ x 14″

89   *Aechmea lamarchei**
Ink; 19½ x 17″

Gifts of the artist

89

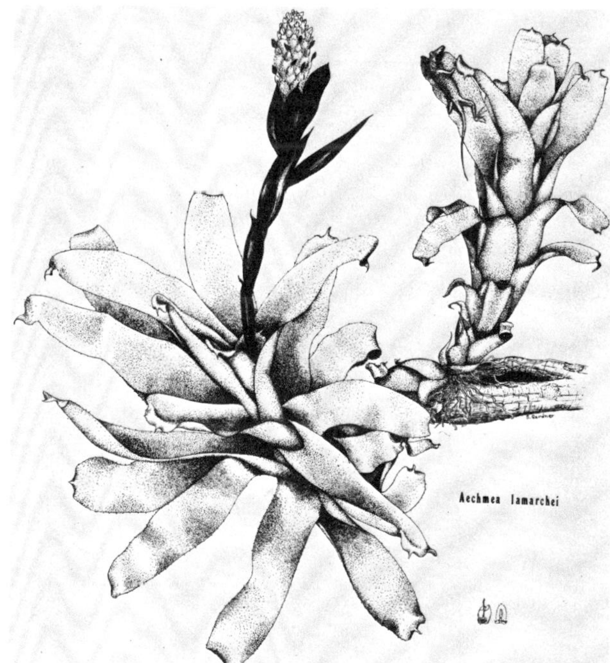

GERRARD, Marguerite Primrose (Mrs. James H.)

Born   Jamaica, West Indies, 29 July 1922. Has lived in the United States since 1959.

Address   15 Colfax Road, Havertown, Pennsylvania 19083.

Education   Royal Drawing Society, London: art, 1938.
Correspondence course in textile design from Harrow, England.

Career   Free-lance artist. Formerly, draughtswoman in charge of the Topographical Drafting Department of Jamaica Bauites, Ltd., subsidiary of the Aluminum Company of Canada.

Media   Tempera, watercolor, oil.

One-person Exhibitions   Booth School, Rosemont, Pennsylvania; The Grange Estate, Havertown; Havertown Library.

Group Exhibitions   Newman Gallery; Pennsylvania Horticultural Society; Chadds Ford, Pennsylvania; Philadelphia Civic Center; Hunt Institute International, 1972; Hunt Institute Travel Shows (Lilies); County Museum, Haverfordwest, Wales; Pennock Florist Inc., Philadelphia; Philadelphia Sheraton Hotel; Fidelity Bank, Independence Mall, Philadelphia.

Commission/Works Published in   Kane, T.R. *Analytical elements of mechanics.* Vol. II. New York and London, Academic Press, 1961.
Christmas card, Pennsylvania Horticultural Society, 1968.

90 Sunflowers, *Helianthus* sp.
Gouache; 17½ x 29⅝″
Lent by the artist

91 Peony, *Paeonia* sp.
Gouache; 21⅞ x 17⅜″

91

GHISLAIN, Raphael Henri-Charles

Born   Uccle, Brabant, Belgium,
9 September 1928.

Address   Heirbaan 13,
2758 Haasdonk, Belgium.

Media   Ink, chalk, wash,
watercolor, oil, etching.

Education   Academy of Fine Arts,
Brussels and Antwerp: 1947-50.
Higher Academy of Fine Arts,
Antwerp: 1950-54.

Career   Painter and draughtsman.

One-person Exhibitions   Antwerp and Brussels.

Group Exhibitions   Hunt Institute International, 1973; Hunt
Institute Travel Shows (International).

Collections   Brussels Cabinet des Estampes; many private
collections in Belgium, France, Holland, and England.

Works Published in   *Six flowers and an arboretum* (calendar).
Antwerp, de Schutter S.A., 1976.
Illustrations for a children's encyclopedia, published in
Belgium (1960-69).

92 Animated Oat, *Avena sterilis*
Watercolor; 26¼ x 17⅛"
Lent by the artist

GOLTE-BECHTLE, Marianne (Mrs. Rolf Heinz Golte)

Born   Frankfurt am Main, Germany, 24 April 1941.

Address   Alexanderstrasse 27, 7 Stuttgart, Germany.

Education   Senckenberg Museum, Frankfurt: 1957-60. Werkkunst-Schule, Wiesbaden, 1960-65.

Career   Botanical and zoological illustrator, Kosmos-Verlag, Stuttgart, 1966 to present.

Media   Watercolor, tempera, pastel, ink, oil crayon, woodcut, linocut.

Group Exhibitions   Hunt Institute International, 1972; Hunt Institute Travel Shows (International); Hunterdon Art Center, Clinton, New Jersey, 1977.

Award   First Prize, Hunterdon Art Center, 1977.

Collection   Kosmos-Verlag, Stuttgart.

Works Published in   Schwammberger, K. *Kleine Säugetiere—Richtig Gepflegt.* Stuttgart, Kosmos-Verlag, 1968.
Kosmos, 12:1968, and 4:1969.
Thenius, E. *Paläontologie.* Stuttgart, Kosmos-Verlag, 1970.
Kastle, W. *Eschen im Terrarium.* Stuttgart, Kosmos-Verlag, 1972.
Aichele, D. *Was blüht denn da?* Stuttgart, Kosmos-Verlag, 1973.   (Also Dutch, Danish, French, and English editions.)
Flauaus, G. *Zwergkaninchen.* Stuttgart, Kosmos-Verlag, 1974.

93 *Eryngium* sp.
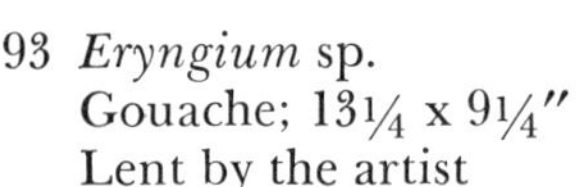
   Gouache; 13¼ x 9¼″
   Lent by the artist

GORDON, John

Born   St. Joseph, Missouri, 17 August 1923.

Address   305 Call Building, Sioux City, Iowa 51101.

Education   Fine Arts School, Marseille, France: 1945; Biarritz-American University, France. Center College, Danville, Kentucky: B.A., University of Iowa, Iowa City: M.F.A., 1951.

Career   Free-lance artist, 1975 to present. Formerly, art teacher at: East Carolina University, Greenville, 1952-60; North Carolina College, 1962-64; Drake University, Des Moines, 1964-67; Westmar College, Le Mars, Iowa, 1972-74. Artist-in-Residence, Des Moines Art Center, 1967-71.

Media   Painting, serigraphy.

One-person Exhibitions   Drake University; University of North Carolina; Duke University; McNider Museum of Art, Mason City, Iowa; Sioux City Art Center; Des Moines Art Center, 1968, 1971.

Group Exhibitions   Numerous, including: Tenth Midwest Annual, Joslyn Art Museum, Omaha, Nebraska; Weatherspoon Annual Drawing Exhibition, University of North Carolina, Greensboro; Rochester Art Center, Minnesota; State Capitol Building, Des Moines, Iowa; Cedar Rapids Art Center.

Awards   Purchase Prize, Annual Exhibition for North Carolina Artists, North Carolina Museum of Art; Purchase Prize, Annual Exhibition for Artists of Virginia and North Carolina, Norfolk Museum of Art; Third Prize, Southeastern Annual, Atlanta Art Association; Purchase Prize, Butler Annual, Butler Institute of American Art. All-Jury Commendation, Midwest Biennial, Joslyn Museum of Art.

Collections   Des Moines Art Center; Sioux City Art Center; Butler Museum of American Art, Youngstown, Ohio; Norfolk Museum of Arts and Sciences; Omaha Art Association; Dubuque Art Association.

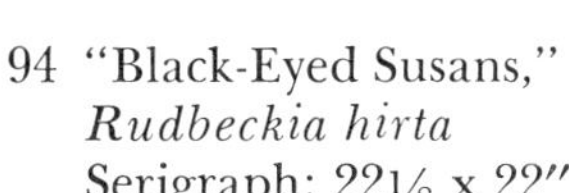

94 "Black-Eyed Susans,"
Rudbeckia hirta
Serigraph; 22⅛ x 22"

GRIERSON, Mary Anderson

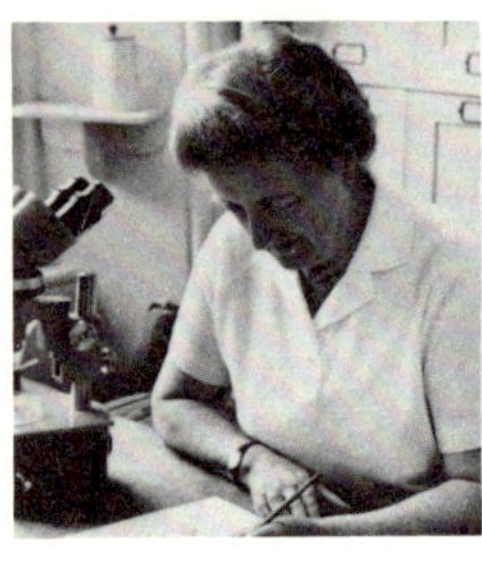

Born   Bangor, North Wales,
27 September 1912.

Address   20 Charmouth Court,
Kings Road, Richmond, Surrey,
England.

Education   Bangor County School
for Girls. Studied painting
under John Nash, East Anglia,
1959-61.

Career   Free-lance botanical artist.
1940 to the present. Recently
specializing in endangered species. Formerly, aerial
photography interpreter, military service, 1940-45.
Cartographical draughtswoman, Hunting Surveys Ltd.,
London, 1946-60. Staff artist, Royal Botanic Gardens, Kew,
1960-72.

Media   Watercolor, ink.

One-person Exhibitions   Hunt Institute, 1975; Pacific Tropical
Botanical Gardens, Hawaii, 1976.

Group Exhibitions   London: The Royal Society, The Linnean
Society, British Museum, Royal Horticultural Society.
International Exhibitions of Botanical Art, Johannesburg;
Royal Botanic Gardens, Kew; Royal Botanic Gardens,
Edinburgh; Frans Hals Museum, Netherlands; Hunt Institute
Internationals, 1968, 1972; Hunt Institute, Artists from Kew,
1974; Hunt Institute Travel Shows (International).

Honors/Awards   Fellow, Linnean Society, London, 1967; Royal
Horticultural Society, London, Gold Medals, 1966, 1969,
1973; Member, Art Workers Guild, London, 1974.

Collections   Royal Botanic Gardens, Kew; Frans Hals Museum,
Haarlem; Society for the Promotion of Nature Reserves,
Lincoln; National Trust, London; World Wildlife Fund,
London; Israeli Nature Reserves Authority; Pacific Tropical
Botanical Gardens, Hawaii; Los Angeles County Museum of
Natural History; many private collections.

Commissions/Works Published in   *British Flora*. Stamp issue
for General Post Office, London, 1967.
Huxley, A.J. *Mountain flowers*. London, Blandford Press,
1967.

Bean, W.J. *Trees and shrubs of the British Isles*. London,
John Murray, Vols. I-IV, 1970-75.
Hunt, P.F. *Orchidaceae*. Bourton, The Bourton Press, 1974.
Numerous issues of *Curtis's botanical magazine, Kew bulletin,
Journal of the Linnean Society, Hooker's Icones Plantarum,
Flora of East Tropical Africa, Flora of Ceylon, Flora
Zambesiaca, Flora of Iraq*.
Illustrations for the Council of Nature, World Wildlife Fund,
National Trust of Great Britain, Royal Horticultural
Society. *
Threatened plants of the world. World Wildlife Fund
[in progress].
Protected plants of Israel. Israeli Nature Reserves Authority
[in progress].
Native flora of the Hawaiian Islands. Pacific Tropical
Botanical Gardens [in progress].
Species tulips. Van Tubergen Nurseries, Holland
[in progress].

95 Common Foxglove, *Digitalis purpurea*
Watercolor; 21½ x 12½″

96 Orchid, *Oncidium ampliatum*
Watercolor and ink; 20½ x 12″

97 "Limestone Flowers of the
Gower Peninsula, South Wales" *
Watercolor; 9 x 12″

Gifts of the artist

95

GROSCH, Laura

Born   Worcester, Massachusetts, 1 April 1945.

Address   P.O. Box, 2495, Davidson, North Carolina 28036.

Education   Wellesley College, Wellesley, Massachusetts: B.A., Art History, 1967. University of Pennsylvania, Philadelphia: B.F.A., Painting, 1968.

Career   Painter and printmaker.

Media   Acrylic, lithography, color pencil.

One-person Exhibitions   Stowe Gallery, Davidson College, Davidson, North Carolina, 1971, 1973; Gallery 501, The Mint Gallery, Charlotte, North Carolina, 1974.

Group Exhibitions   Numerous, including: Impressions Gallery, Boston, 1973; North Carolina Museum of Art, Raleigh, 1973; Colorprint USA, Texas Technical University, Lubbock, 1973-74; New American Graphics, Madison Art Center, Wisconsin, 1975; Plants and Flowers, Roanoke Fine Arts Center, 1975; 11th International of Graphic Art, Moderna Galerija, Ljubjlana, Yugoslavia, 1975; Artists' Biennial, New Orleans Museum of Art, 1975; National Invitational, Art Center, Winston Salem, North Carolina, 1977; Art U.S.A.: The South, U.S. Information Agency, touring Central and South America, 1976-77; 30 Years of American Printmaking (including 20th National Print Exhibition), Brooklyn Museum, 1976-77.

Honors/Awards   Purchase Award, 5th Biennial International Mat Media Exhibition, Dickinson State College, Dickinson, North Dakota, 1975; Purchase Award, Western Annual Juried Exhibition of Works on Paper, Western Illinois University, Macomb, 1975; First National Bank of Boston Purchase Award, 27th Annual Boston Printmakers Exhibition, Museum of Fine Arts, Boston, 1975; Purchase Award, 10th Annual Piedmont Graphics Exhibition, Mint Museum of Art, Charlotte, North Carolina, 1973.

Collections   British Museum, London; Victoria and Albert Museum, London; California Palace of the Legion of Honor; New York Public Library; Smithsonian Institution; Bowdoin College, Maine; Museum of Fine Arts, Springfield, Massachusetts; Dayton Art Institute; Museum of Fine Arts, Boston; Minneapolis Institute for the Arts; Greenville Museum of Art, South Carolina; Library of Congress, Washington, D.C.; The Ringling Museum, Sarsota, Florida.

98 Asparagus
Color lithograph; 29½ x 21½″

99 Amaryllis
Color lithograph; 29½ x 21⅞″

99

GRUTER, Adrianus (Ad.)

Born   Utrecht, Netherlands,
30 August 1946.

Address   Nieuwegein, Ereprijs 3,
Netherlands.

Education   Academy of Art,
Artibus, Utrecht: studied
publicity and sculpture, final
examination, 1968.

Career   Free-lance illustrator and
jazz musician. Formerly, sculptor,
St. Stevens Church, Nijmegen,
1968-69. Botanical artist, State Agricultural University,
Wageningen, 1969-70. Scientific illustrator, Laboratorium
Veterinaire Anatomie, Utrecht, 1970-73.

Medium   Ink.

Works Published in   Dyce, U.M. *Essentials of bovine
anatomy.* Utrecht, 1971.
Dyce, U.M. *Inleiding tot de anatomie.* Utrecht, 1972.
75 years of botanical gardens. Wageningen, 1972.
Breteler, F.J. *The African Dichapetalaceae.* Wageningen,
1973. *
Hartman, W. *The pelvic outlet in female goats.* Utrecht,
1973.
Hartman, W. *Digestietractus.* Utrecht, 1974.
Jansen, L.H. *Huid en geslachtsziehten.* Utrecht, 1975.
(with Uingma, M.J.) *Nederlands leerboek de orthopedie.*
Utrecht, 1975.
Lohman, A.H.M. *Anatomie voor de tandarst.* Utrecht
[in progress].
Menuhin, H. *Exercises for the musician.* Utrecht
[in progress].

100 *Dichapetalum* sp.*
Ink; 8⅜ x 5¾″
Permanent loan of the Laboratorium voor
Plantensystematiek en -Geografie,
Landbouwhogeschool, Wageningen, Netherlands

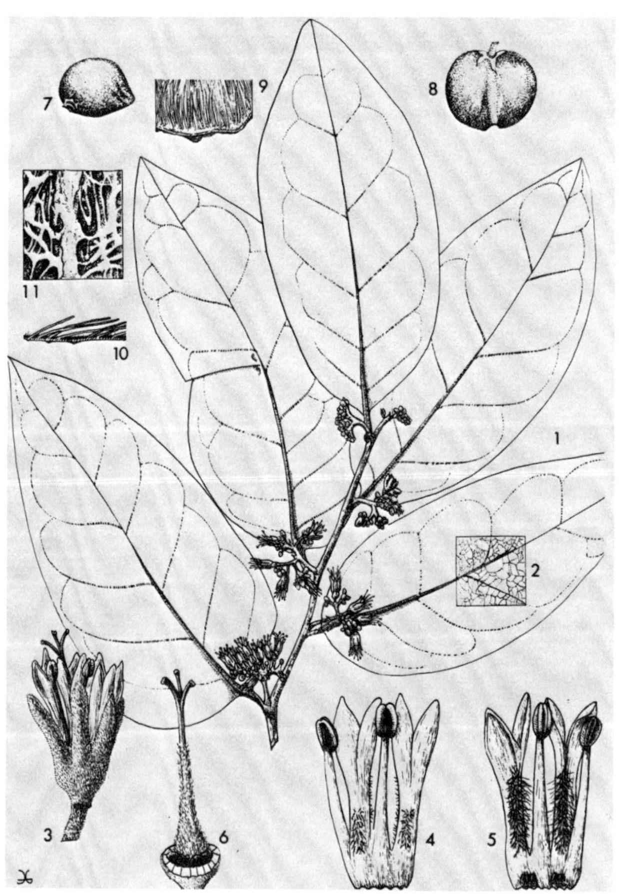

HAMMOND, Catherine R. (Mrs. Franklin T. Hammond, Jr.)

Born   Brookline, Massachusetts, 8 January 1909.

Died   November 1976.

Education   Amy Sacker School of Design, Boston: 1930. Boston Museum of Fine Arts: 1955-59. Studied graphics, 1960.

Career   Free-lance artist and wildflower illustrator.

Media   Watercolor, pencil, ink; portraits in conté crayon and oil.

One-person Exhibitions   In Massachusetts: Cambridge Art Association; Boston Public Library, 1968; Thoreau Center, Concord, 1973; Arnold Arboretum, 1974; Northeast Wild Flower Society, Horticulture Hall, Boston, and Garden-in-the-Woods, Framingham; Newton Public Library; Plymouth Public Library; Cambridge Center for Adult Education; Edna Stebbins Gallery; First Parish Church, Cambridge.

Collection   "The Vale," Waltham, Massachusetts.

Commissions/Works Published in   Taylor, K.S. and Hamblin, S.S. *Handbook of wild flower cultivation, a guide to wild flower cultivation in the home garden.* New York, Macmillan Company, 1963.

Newcomb, L. *Pocket key to common wild flowers.* (Cover design) Framingham, Massachusetts, New England Wild Flower Society, 1963.

Gibbons, E. *Stalking the blue-eyed scallop.* David McKay Company, 1963.

Coffin, G. and Lewis, M. *Twenty common mushrooms and how to cook them.* Boston, Bruce Humphries, International Pocket Library, 1965.

Hammond, C. "A gallery of herbs," *Horticulture,* March 1976.

Wall charts for New England Wild Flower Society, Framingham, Massachusetts: *Poisonous plants of the North East, native and introduced,* 1961; *Twenty-five common mushrooms of New England,* 1967;*Twenty-five summer wild flowers of roadsides and fields in New England,* 1974.

101 Jimson-Weed, *Datura stramonium**
Watercolor; 8¼ x 6¼"

102 Jack-in-the-Pulpit, *Arisaema triphyllum**
Watercolor; 8¼ x 6¼"

103 Common Foxglove, *Digitalis purpurea**
Watercolor; 9¼ x 6⅛"

104 White Baneberry, *Actaea alba**
Watercolor; 9 x 6"

Gifts of the artist

102

HART, Nancy Shinn (Mrs. Richard)

Born   Chicago, Illinois, 28 August 1940.

Address   331 Union Avenue, Batavia, Illinois 60510.

Education   Art Institute of Chicago: graduated 1962.

Career   Head of Graphic Arts Department, Morton Arboretum, Lisle, Illinois, January 1977 to present. Formerly, artist, exhibit designer, and Curator of Prints and Drawings, Morton Arboretum, 1966-1977. Instructor of botanical art and illustration, Fernwood, Inc., Niles, Michigan (6 years).

Media   Ink, oil, watercolor, etching, lithography, sculpture.

One-person Exhibitions   Evanston Community Center, Illinois, 1962; Oak Park High School, Illinois, 1963; Morton Arboretum, 1967.

Group Exhibitions   Hunt Institute International ,1972; Art Institute of Chicago; Fermi National Accelerator Laboratory.

Collections   Morton Arboretum; Encyclopaedia Britannica; many private collections.

Commissions/Works Published in   *Compton's encyclopedia.*
The Encyclopaedia britannica.
Sierra Club publications.
Morton Arboretum quarterly, and other Arboretum publications. *
Illustrations for Audubon Society and Nature Conservancy, Chicago.

105 Mountain Rose Bay,*
Rhododendron catawbiense
Ink; 11⅜ x 17⅝″
Lent by The Morton Arboretum

HARVEY, Norman Bruce

Born   Palmerston North, New Zealand, 28 July 1931.

Address   93 Kauri Point Road, Laingholm, Aukland 7, New Zealand.

Education   Timaru Boys High School. Otago University. Australian Institute of Advertising: L.A.I., 1956.

Career   Botanical and wildlife painter. Formerly, copy chief and copy visualizer with Curzon Ferri Thompson Advertising Ltd., Aukland. Has also worked in radio and newspaper industries. Author of two novels. Has recently embarked on a new art form called "Harvonic," which combines music, blank verse, and painting in one medium.

Media   Gouache, tempera, pencil, ink, oil.

One-person Exhibitions   John Leech Gallery, Aukland, 1969; Rothmans Cultural Foundation, 1971; Kennedy Galleries, New York, 1975; Museum of Natural History, New York, 1975; Gallery Hawaii, Honolulu, 1976; Nottingham City Art Gallery, 1976.

Group Exhibitions   Hunt Institute International, 1972; Hunt Institute Travel Shows (International).

Award   Novel of the Year Award, New Zealand Literary Foundation, 1966.

Collections   Mrs. Lyndon Baines Johnson; The Honorable George Aryoshi, Governor of Hawaii; United States Embassy, New Zealand.

Commissions/Works Published in   Harvey, N.B. and Godley, E.J. *New Zealand botanical paintings.* Christchurch, Whitcombe and Tombs, Ltd., 1968.
Harvey, N.B. *A portfolio of New Zealand birds.* Wellington, A.H. & A.W. Reed, 1970.
Weekly news, Women's weekly, Eve, Company director, New Zealand heritage, New Zealand arts horizon.

106 Grapefruit, *Citrus paradisi*
    Gouache; 15½ x 25¼"

HAYES, Marvin Everett

Born   Canton, Mississippi, 30 September 1939.

Address   14 Wakefield Road, Wilton, Connecticut 06097.

Education   Texas A & M University, College Station: 1958-59. Lamar University, Beaumont, Texas: B.F.A., 1963. Art Students League, New York: 1964. Columbia University, New York: M.F.A., 1966.

Career   Free-lance artist.

Media   Egg tempera, etching.

One-person Exhibition   FAR Gallery, New York, 1976.

Group Exhibitions   FAR Gallery, 1971 to present.

Award   First Prize, New England Annual, Silvermine Guild, New Canaan, Connecticut, 1973.

Collections   Pennsylvania State University; Westmoreland County Museum of Art, Greensburg, Pennsylvania; Museum of Modern Art, New York; National Portrait Gallery, London; Rijksmuseum, Netherlands; Vatican, Rome; Musée d'Art Moderne, Bibliothèque Nationale, and Louvre, Paris.

Commissions/Works Published in   Numerous journals, including: *Playboy, McCalls, Red book, Esquire, Reader's digest, Ladies' home journal, Boys' life, Cosmopolitan, Seventeen, American artist, Southern living, Argosy.*
Books published by: Ballentine Books; Bantam Books; Dell Publications; Doubleday; Macmillan Publishing Co., Inc.; Avon Books.
Dickey, James. *God's Images,* Birmingham, Oxmoor house [in progress].
Illustrations for MGM Records.

107 Oak, *Quercus* sp.
Etching; 23½ x 14⅞″
Gift of the artist

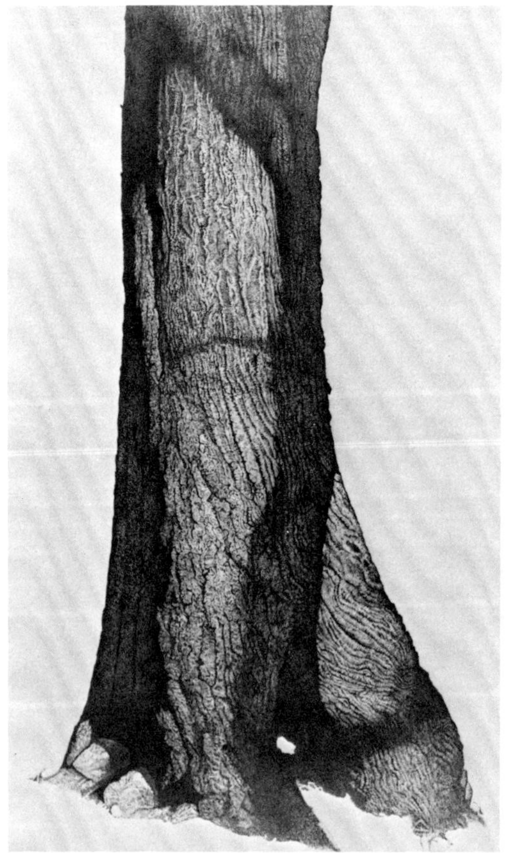

HEINS, Esther (Mrs. Harold)

Born   Brooklyn, New York, 10 November 1908.

Address   8 Mitchell Road, Marblehead, Massachusetts 01945.

Education   Massachusetts College of Art: B.S., Education, 1929. Studied oil painting with Ernest L. Major. Studied with William McNulty, Ann Brockman, and John Corbino, summer courses, Cape Ann School, Rockport, Massachusetts. Studied with Oskar Kokoschka, School of Vision, Salzburg, Austria, 1956.

Career   Free-lance botanical artist. Formerly, free-lance artist for the leading department stores in Boston. Illustrated children's school books for D.C. Heath & Co.

Media   Watercolor, oil, pastel.

One-person Exhibitions   Boston Public Library, 1966, 1970; Rockport (Massachusetts) Art Association, 1969; Hilles Library, Radcliffe College, Cambridge, 1976.

Group Exhibitions   Hunt Institute International, 1972; Hunt Institute Travel Shows (International); Hunterdon Art Center, Clinton, New Jersey, 1977.

Commissions/Works Published in   *Boston Public Library news*, 1966, 1970. *Harvard University Gazette*, 1976. *Flowering trees and shrubs of the Arnold Arboretum* [in progress].

108   *Anthurium andraeanum*
Watercolor; 29½ x 21½"

109   *Ampelopsis brevipedunculata* var. *elegans*
Watercolor; 18½ x 13¾"

108

HELMER, Jean Cassels

Born   Nashville, Tennessee,
25 December 1941.

Address   726 South Boulevard,
Evanstown, Illinois 60202.

Education   Eastern Michigan
University: 1960-62. Art Institute
of Chicago: 1964-67.

Career   Free-lance illustrator.

Media   Ink, pencil, acrylic.

Commissions/Works Published in
*World book encyclopedia, Encyclopaedia britannica,
Science year, Childcraft annual,* and other trade, education,
and advertising books. Publishers include A.B. Morse,
Scott Forseman, Rand McNally, Silver Burdett, Follett,
and Harper & Row.

110   Wood-Sorrel, *Oxalis* sp.
Ink; 10 x 13″

111   Apple, *Malus sylvestris*
Ink; 9⅝ x 7½″

112   Barberry blossom, *Berberis* sp.
Ink; 3½ x 4¾″

113   Mallow family, Malvaceae
Ink; 6½ x 4½″

All lent by the artist

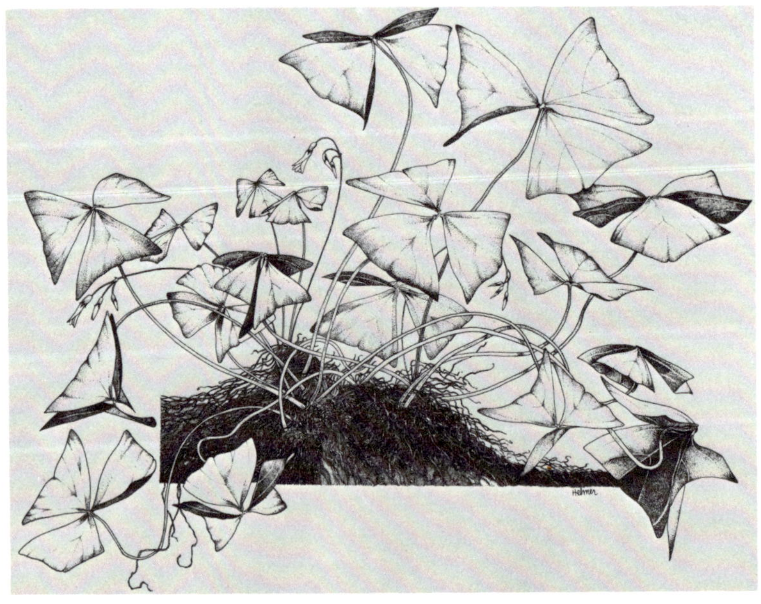

110

HENNESSY, Esmé Frances Franklin

Born   Umzinto, Natal, South Africa, 30 August 1933.

Address   29 Venice Road, Durban 4001, Natal, South Africa.

Education   Natal University: B.Sc., botany and zoology, 1956; B.Sc., botany, 1958; M.Sc., botany, 1966.

Career   Senior Lecturer, Botany Department, University of Durban-Westville, Natal, 1972 to present. Angiosperm taxonomist, with interest in pteridophyte and gymnosperm taxonomy and morphology. Formerly, Research Assistant, Amoebiasis Research Unit, Institute for Parasitology, Council for Scientific and Industrial Research, 1958-60. Lecturer, University College, Durban, 1961-72. Commemorated in *Erythrina × hennessyae* Barneby and Krukoff. Signs botanical artworks "E.F. Hennessy" and decorative works "Frances Franklin."

Media   Watercolor, ink.

Group Exhibitions   Tatham Art Gallery, Pietermaritzburg, 1966; Pieter Wenning Gallery, Johannesburg, 1969; International Exhibition of Botanical Art, Johannesburg, 1973; South African Association of Arts Gallery, Pretoria, 1974; Exotica 74, Durban, 1974.

Collections   Standard Bank of South Africa, Johannesburg and Durban; Botanical Research Institute, Pretoria; Roberts Construction, Johannesburg; Basil Read Engineering, Johannesburg; Department of Foreign Affairs, Republic of South Africa; many private collections.

114 Sausage Tree, *Kigelia africana*
Watercolor; 27¾ x 20¼"
Lent by the artist

115 African Tulip Tree, *Spathodea campanulata*
Watercolor; 26½ x 18½"
Lent by the artist

116 Senna, *Cassia didymobotrya* (top),
Coral Tree, *Erythrina caffra*
Watercolor; 13¼ x 10½"
Gift of the artist

117 Lion's-ear, *Leonotis leonurus* (top right),
Wild Cineraria, *Senecio macroglossus* (top left),
Coral Tree, *Erythrina lysistemon* (center),
Barleria repens (bottom)
Watercolor; 14 x 10"
Gift of the artist

Works Published in   Wilmot, A.J. *Clinical Amoebiasis.* Oxford, Blackwell, 1962.

Crass, R.S. *Freshwater fishes of Natal.* Pietermaritzburg, Shuter and Shooter, 1964.

Hennessy, E.F. "Six common wild flowers," *Veld & flora,* March, 1971.

Hennessy, E.F. *South African Erythrinas.* Durban, Wild Life Protection and Conservation Society, 1972.

Hennessy, E.F. "Three pepos," *Veld & flora,* March, 1972. *Journal of the Botanical Society of South Africa,* 1972.

Hennessy, E.F. "Three pentamerous flowers," *Veld & flora,* March, 1973.

Hennessy, E.F. "Erythrina caffra," *Flowering plants of Africa,* 1976.

Hennessy, E.F. "Erythrina latissima," *Flowering plants of Africa,* 1976.

Hennessy, E.F. "Erythrina in southern Africa," *Journal of the Botanical Society of South Africa,* 1976.

Wild fruits of South Africa. Suite of paintings reproduced by Howard Swan, Johannesburg, 1976.

114

HERKLOTS, Geoffrey Alton Craig

Born   Naini Tal, India, 10 August 1902.

Address   Vanners, Chobham, Woling, Surrey GU24 8SJ, England.

Education   Trent College, Derbyshire; M.Sc. University of Leeds: Ph.D., Trinity Hall, Cambridge.

Career   Biologist. Formerly, editor, *Hong Kong naturalist*, 1930-41. Reader in biology, University of Hong Kong, 1928-45. Secretary for Development, Hong Kong, 1946-48. Secretary for Colonial Agriculture Research, Colonial Office, London, 1948-53. Principal and Director of Research, Imperial College of Tropical Agriculture, Trinidad, 1953-60. Retired, 1961. Colombo Plan Botanical Adviser to HM Government of Nepal, 1961-63.

Medium   Ink.

Honor   Fellow, Linnean Society.

Works Published in   *Common marine food fishes of Hong Kong.* 1936, 1940.
Journal of Hong Kong Fisheries Research Station, 1940.
Vegetable cultivation in Hong Kong. 1941, 1947.
The birds of Hong Kong, field identification and field note book. 1946, 1952.
The Hong Kong countryside. 1951.
Hong Kong birds. 1953.
Birds of Trinidad and Tobago. 1961.
Vegetable cultivation in South-East Asia. 1973.
Flowering tropical climbers. 1976. *

118 *Agapetes macrantha**
Ink; 13¼ x 10¾″

119 Morning-Glory, *Ipomoea horsfalliae* *
Ink; 9⅞ x 8″

120 Passion-Flower, *Passiflora racemosa**
Ink; 14⅞ x 6¾″

Gifts of the artist

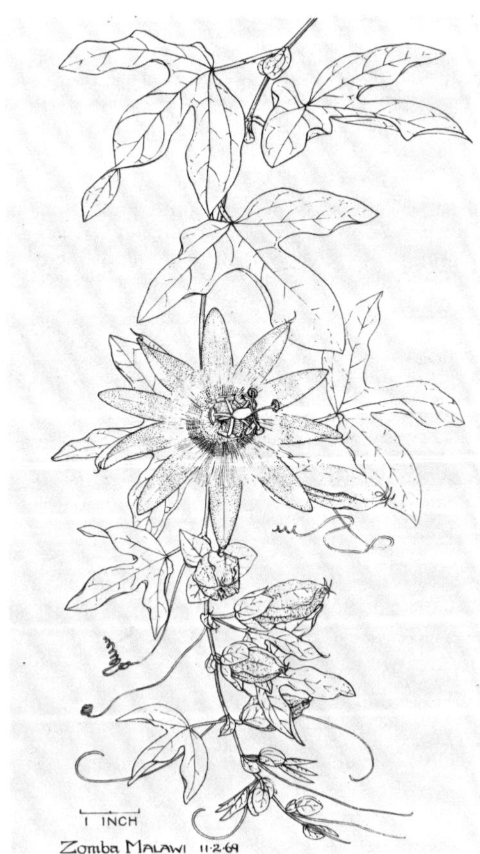

120

HILHORST, Gerhardus P. L.

Born   Hillegersberg, Netherlands, 19 March 1886.

Died   1957.

Education   State School of Industrial Art, Amsterdam: diploma in drawing and perspective, 1905.

Career   Fine art instructor, Java, Indonesia, 1920-39. Formerly, draughtsman and designer of metalware, Royal Metalware Factory of Daalderop & Son, Tiel; and designer of book decorations, diplomas, seals, etc., for Art Press, Mouton & Co., The Hague, prior to 1910. Utrecht University: draughtsman, Faculty of Surgery; drawing and anatomy instructor, School of Fashion Design, 1910-14. Drawing instructor, School of Graphic Design, Utrecht University, 1914-20.

Media   Woodcut, watercolor, pencil, silverpoint.

Group Exhibition   Zeist, Netherlands, 1947.

Collections   Private collections in the Netherlands and South Africa.

Works Published in   Various issues of *Boekcier*, Netherlands.

121 Field Flowers
     Silverpoint; 8⅛ x 6¼″

122 Snapdragon, *Antirrhinum majus*
     Silverpoint; 7½ x 7⅛″

Gifts of Ellaphie Ward-Hilhorst

121

HNIZDOVSKY, Jacques

Born   Pylypcze, Ukraine, 27 January 1915. Has lived in the United States since 1949.

Address   5270 Post Road, Riverdale, New York 10471.

Education   Academy of Fine Arts, Zagreb, Yugoslavia: Diploma, 1942.

Career   Self-employed artist.

Media   Woodcut, oil, pencil, etching, sculpture, ceramics.

One-person Exhibitions   More than 50 since 1954, including: Pratt Institute, New York; Associated American Artists, New York; Lumley-Cazalet Gallery, London; Winnipeg Art Gallery; Tahir Gallery, New Orleans; Long Beach Museum of Art, California, 1977.

Group Exhibitions   Numerous, including: American Graphic Arts (Tokyo and U.S.S.R.); United States Information Agency (Europe, Asia, Africa, and South America); [2nd] Triennale Internazionale della Xilografia Contemporanea, Capri; University of Illinois.

Awards   Minneapolis Institute of Art, 1950; Associated American Artists, 1959, 1962; Tiffany Fellowship, 1961; Museum of Fine Arts, Boston, 1962; MacDowell Colony Fellowship, 1963, 1976-77.

Collections   Numerous, including: Philadelphia Museum of Art; Library of Congress and National Collection of Fine Arts, Washington, D.C.; Museum of Fine Arts, Boston; Cleveland Museum of Art; Butler Institute of American Art, Youngstown, Ohio; Nelson Rockefeller; many private collections.

Works Published in   Graff, M.M. *Three trails in Central Park.* New York, Greensward Foundation, Inc., 1970.

DeWolf, G. *Flora exotica.* Boston, David R. Godine, 1972.

Rannit, A. "Jacques Hnizdovsky, A Ukrainian Graphic Artist," *The Yale University Library gazette,* Vol. 51, No. 4, April 1977.

Tahir, A. *Hnizdovsky—Woodcuts, 1944-1975.* Gretna, Louisiana Pelican Publishing Company, 1976.

Sheep in wood. The Story of a woodcut with Jacques Hnizdovsky (Documentary film). New York, Slavko Nowytski, 1971.

123   "Syngonium 1973"
Woodcut; 27½ x 10"

124   "Louisiana Champion Live Oak,"
Quercus virginiana
Woodcut; 15¾ x 24½"
Lent by the artist

125   "Fern"
Woodcut; 15½ x 21½"
Lent by the artist

124

HOEK, Adriana Elisabeth Annelies

Born   Noordwijk aan Zee, Netherlands, 13 August 1951.

Address   Broekstraat 377, Arnhem, Netherlands.

Education   Academy of Art, Arnhem: studied drawing, painting, graphic art and photography. Studied illustration with Bert Bouman. Studied graphic art with Harry V. Kruiningen and E. Maliangkay.

Career   Free-lance illustrator (children's books) and photographer; teaches creative development and crafts to children. Formerly, botanical illustrator, State Agricultural University, Wageningen, 1971.

Media   Ink, etching, oil, watercolor.

Group Exhibitions   Academy of Art, Arnhem, 1974; Stedelijk Museum, Amsterdam, 1974.

Works Published in   van der Maessen, L.J.G. *Cicer L., a monograph of the genus with special reference to the chickpea (Cicer arietinum L.), its ecology and cultivation.* Wageningen, 1972.*
Start '74. Steendrukkerij de Yong & Company, 1974.
Jaarverslag van de Nederlandse GAS UNIE, 1974.
Werkgroep Brinkkemper. *Dat less ik 6 sniep, snap en snoep.* Zutphen, Thieme, 1975.

126 *Cicer pinnatifidum**
   Ink; 8 x 5½"

127 *Cicer bijugans **
   Ink; 8 x 5½"

Permanent loans of the Laboratorium voor Plantensystematiek en -Geografie, Landbouwhogeschool, Wageningen, Netherlands

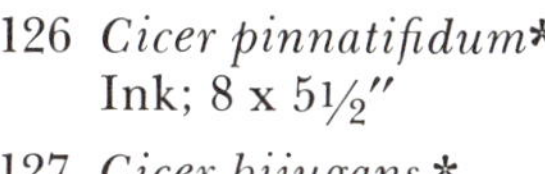

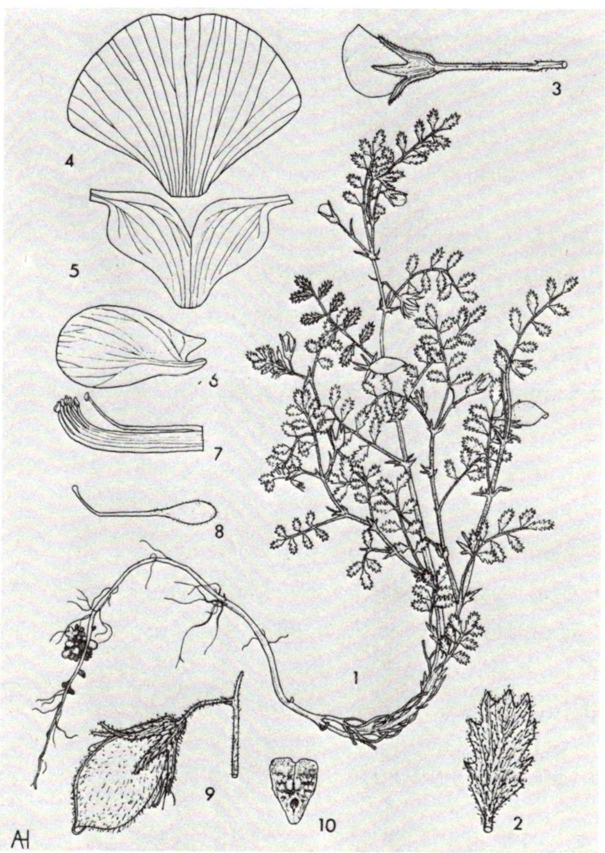

126

HOLMQVIST, Kristian

Born   Lund, Sweden, 5 June 1928.

Address   Hjärupsv. 28, 22248 Lund, Sweden.

Education   Katedralskolan, Lund: 1940-47. Skånska målarskolan, Malmö: studied painting, 1947-48. Folkskoleseminariet, Lund: Degree, Elementary School Teacher, 1952. Essem-skolan, Malmö: studied painting, 1953. Studied drawing and sculpture in Malmö and Lund.

Career   Teacher and illustrator.

Media   Pencil, ink, watercolor.

One-person Exhibition   Smålands Konstarkiv, Värnamo, 1975.

Group Exhibitions   Sverigehuset, Stockholm, 1977; Hunterdon Art Center, Clinton, New Jersey, 1977.

Awards   Stipendium Sveriges Författarfond, 1974, 1975.

Works Published in   Sjögren, B. *Träd i svenska marker (Swedish trees)*. Göteborg, 1962.
Sjögren, B. *Småkryp i svenska marker (Swedish insects)*. Göteborg, 1963.
Andersson-Malmberg. *Möte med växter (A school book about plants)*. Lund, 1963.
Lägnert-Stenervik. *Geografi för grundskolans mellanstadium (A school book on geography)*. Lund, 1965.
Ternby. *Vi lär oss skriva maskin (A handbook of typewriting)*. Malmö, 1969.
Sjögren, B. *Brannvinskoyddor i skog och mark (A book about plants to be used as spices in spirits)*. Stockholm, 1975.

128 Horse-Chestnut, *Aesculus hippocastanum*

129 Woodbine, *Lonicera periclymenum*

130 Cornelian-Cherry, *Cornus mas*

131 Common Garden Sunflower, *Helianthus annus*

132 Leek, *Allium porrum*

133 Dandelion, *Taraxacum vulgare*

134 Pear, *Pyrus communis*

135 Common Tulip, *Tulipa gesneriana*

136 Maize, *Zea mays*

137 Tomato, *Solanum lycopersicum*

138 Goat Willow, *Salix caprea*

139 Bittersweet, Nightshade, *Solanum dulcamara*

Watercolors; 2¾ x 3⅛"
All lent by the artist

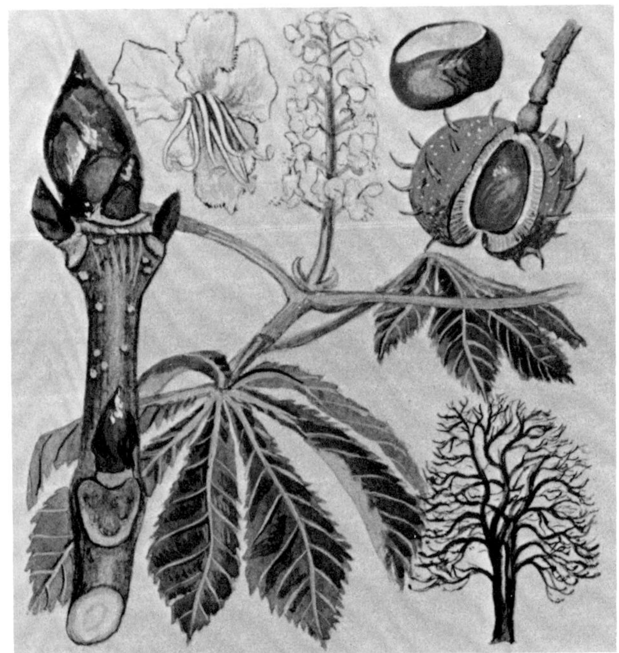

128

HOWE, Marsha Kristen Heinbaugh

Born   Chardon, Ohio, 1 October 1943.

Address   1208 North Corona Street, Colorado Springs, Colorado 80903.

Education   Syracuse University: B.F.A., printmaking/painting, 1968.

Career   Senior designer and illustrator, Looart Press, Inc., Colorado Springs. Owner of Romar Graphics, printmaking business.

Media   Etching, gouache, ink.

One-person Exhibition   George Nix Gallery, Colorado Springs, 1975.

Group Exhibitions   Randi's Art Gallery, Denver; Gallery 323, Casper; Gilpin County Arts Association, Central City, Colorado; The Arts Association, Johnstown, Pennsylvania; The Broadmoor Community Annual Art Show, Colorado Springs; Ford Motor Company Annual Art Show, Dearborn, Michigan.

Honors/Awards   Best Graphic, Garden Valley Art Show, 1973; Best of Show, Ford Motor Company Annual Art Show, 1969.

Collection   Syracuse University School of Art Collection.

Works Published in   Greeting cards produced by Looart-Current, Inc., Gibson Greeting Cards, Inc. and Northlight Publications.

140 "Columbine," *Aquilegia* sp.
   Etching; 10¼ x 4⅞" plate mark

141 "Weeds II"
   Etching; 14⅜ x 9⁵⁄₁₆" plate mark

140

HULL, Gregory

Born   London, England, 13 June 1946.

Address   % Lancaster, 225 East Houston Street, New York, New York 10002.

Education   Bath Academy of Art: National Diploma in Art and Design, 1968. Royal College of Art: M.A., Fine Art, 1971.

Career   Artist.

Medium   Color pencil.

One-person Exhibitions   Clare College Art Gallery, Cambridge, England, 1971; Kornblee Gallery, New York, 1976.

Group Exhibitions   Oxford University, 1969; A.I.A. Gallery, London, 1970; Morehead State University, Morehead, Kentucky, 1969; Garage Art Gallery, London, 1974.

Collection   King's College, Cambridge.

142 Mushroom, *Amanita excelsa*
    Color pencil; 8 x 6¾″

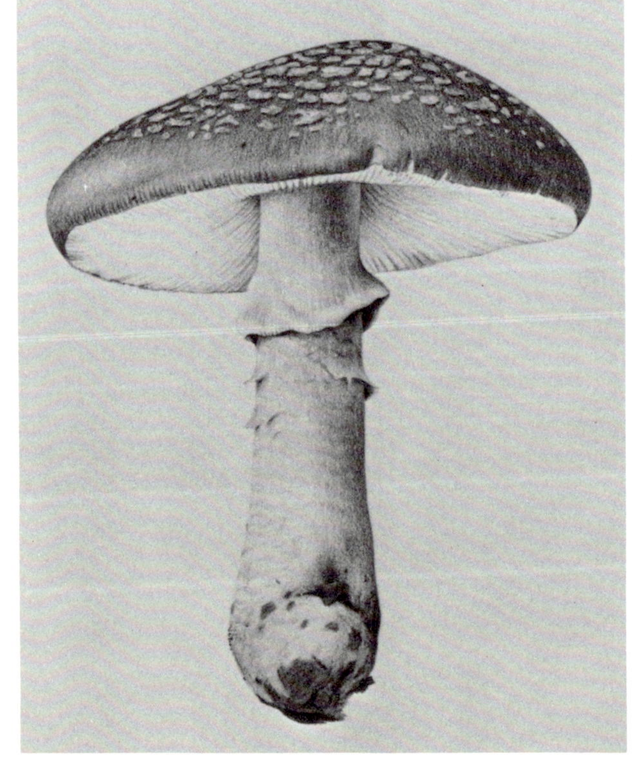

JAMISON, Philip

Born   Philadelphia, Pennsylvania, 3 July 1925.

Address   104 Price Street, Westchester, Pennsylvania 19380.

Education   Philadelphia Museum School of Art: 1946-50.

Career   Painter. Formerly, watercolor instructor, Philadelphia College of Art, 1961-63. NASA artist for Apollo-Soyuz space launch.

Medium   Watercolor.

One-person Exhibitions   Hirschl and Adler Galleries, New York, 1959, 1962, 1965, 1967, 1969, 1971, 1974, 1976; Duke University, 1969; Delaware Art Museum, 1973; Sessler Gallery, Philadelphia, 1963, 1972; Janet Fleisher Gallery, Philadelphia, 1977.

Group Exhibitions   Metropolitan Museum of Art, New York; Philadelphia Museum of Art; Museum of Fine Arts, Boston; Cleveland Institute of Art; Butler Institute of American Art, Youngstown, Ohio; National Academy of Design, New York.

Honors/Awards   First Award, Wilmington Society of the Fine Arts, 1957, 1959, 1961; Bainbridge Award, Allied Artists of America, 1958, 1960; Dawson Medal, Pennsylvania Academy of Fine Arts, 1959; First Award, National Arts Club, New York, 1960; Watercolor Medal of Honor, Knickerbocker Artists, 1961; William Church Osborn Memorial Award, American Watercolor Society, 1961; First Prize, Boardwalk Show, Atlantic City, 1961; Dana Medal, Philadelphia Watercolor Club, 1961; Lena A. Mason Prize, National Academy of Design, 1962; Ranger Fund Purchase, National Academy of Design, 1962; M.W. Zimmerman Prize, Philadelphia Watercolor Club, 1963; Gold Medal of Honor, Allied Artists of America, 1964; Childe Hassam Fund Purchase, American Academy of Arts and Letters, 1965; Lily Saportas Award, American Watercolor Society, 1965; C.F.S. Award, American Watercolor Society, 1966; National Academy of Design, 1967; Thornton-Oakley Memorial Prize, Philadelphia Watercolor Club, 1967; Samuel Finley Breese Morse Medal, National Academy of Design, 1969; Whitney Award, American Watercolor Society, 1971, 1973; High Winds Award Medal, American Watercolor Society, 1972. Elected Academician, National Academy of Design, 1970.

Collections   Pennsylvania Academy of the Fine Arts, Philadelphia; Delaware Art Museum, Wilmington; National Academy of Design, New York.

Commissions/Works Published in   Meyer, S. and Kent, N. *Watercolorists at Work.* Watson-Guptill, 1972. *American Artist.*
Reproductions of watercolors produced by New York Graphic Society, American Artists Group, Inc., Philadelphia Editions Ltd., Aaron Ashley Inc.

143   "Island Daisy Field"
Watercolor; 9¼ x 12⅛"

JULES, Mervin

Born   Baltimore, Maryland, 21 March 1912.

Address   Art Department, City College, Convent Avenue at 138th Street, Room E112, New York, New York 10031.

Education   Baltimore City College: degree, 1930. Maryland Institute of Fine and Applied Arts: degree, 1933. Studied with Thomas Benton, Art Student's League, New York, 1934.

Career   Professor and Chairman of the Art Department, City College, New York, 1969 to present. Formerly, taught art in New York, 1942-44: Fieldston School, Museum of Modern Art, War Veterans Art Center, and People's Art Center. Taught at University of Massachusetts, Extension Division, 1946. Visiting artist, 1945-46, Associate Professor, 1946-63, and Professor, 1964-69, Department of Art, Smith College, Northampton, Massachusetts. Taught at: Highfield Art Workshop, Falmouth, Massachusetts, 1950; University of Wisconsin, Madison, 1951; George Walter Vincent Smith Art Museum, Springfield, Massachusetts, 1953; University of Michigan, Ann Arbor, 1962.

Medium   Woodcut.

One-person Exhibitions   More than 50, including: Hudson D. Walker Gallery, New York, 1937, 1939, 1940; Baltimore Museum of Art, 1940; Weyhe Gallery, New York, 1941; A.C.A. Gallery, New York, (9 shows, 1941-59); Whythe Gallery, Washington, D.C., 1945; Garelick Gallery, Detroit (5 shows, 1945-55); Associated American Artists Gallery, New York, 1961; Fran-Nell Gallery, Tokyo, 1967.

Group Exhibitions   Numerous, including: Corcoran Art Gallery, Washington, D.C.; Whitney Museum and Museum of Modern Art, New York; Pennsylvania Academy, Philadelphia; Museum of Modern Art, Paris; Hunt Institute Internationals, 1968, 1972; Hunt Institute Travel Shows (Botanical Prints).

Honors/Awards   Numerous, including: Fellow, Royal Society of Art, United Kingdom, 1976; Trustee, Cummington School of Music and Art; Trustee, Provincetown Art Association; 125th Anniversary Medal, City College, New York, 1973; Fellow, McDowell Colony, 1938, 1941, 1961. Purchase Prizes: Associated American Artists, 1959, and Brooklyn Museum, 1946.

Collections   More than 40, including: Metropolitan Museum of Art and Museum of Modern Art, New York; Museum of Fine Arts, Boston; Library of Congress, Washington, D.C.; Baltimore Museum of Art; Philadelphia Museum of Art; Brooklyn Museum; Fogg Art Museum, Cambridge; Carnegie Institute, Pittsburgh; Tel Aviv Museum; Swiss Consulate; Joseph Hirshorn.

Works Published in   Print editions, Christmas card designs, and illustrations for children's books and magazines.

144 "Sunflowers," *Helianthus* sp.
Color woodcut; 21⅜ x 17¼"

KLEIN, Karen Anne

Born   Kankakee, Illinois,
28 October 1942.

Address   120 Cambridge, Pleasant
Ridge, Michigan 48069.

Education   University of Michigan,
Ann Arbor: B.S., Design, 1964.
Wayne State University, Detroit:
M.A., 1967.

Career   Painting Instructor, Henry
Ford College, Dearborn,
Michigan, 1976 to present.

Media   Watercolor, etching, woodcut.

One-person Exhibitions   Arnold Klein Gallery, Royal Oak,
Michigan, 1976; Aquinas College, Grand Rapids, Michigan,
1977.

Collections   Kresge World Headquarters, Troy, Michigan;
City National Bank, Detroit; Henry Ford College, Dearborn;
Kalamazoo Institute of Arts, Kalamazoo; Michigan Bell
Telephone Company, Detroit; Bon Secours Hospital, Grosse
Pointe, Michigan.

Works Published in   Print editions.

145 Tulips
Watercolor; 17¼ x 13¾"

146 "Interior with Bouquets"
Watercolor; 24¾ x 18¼"

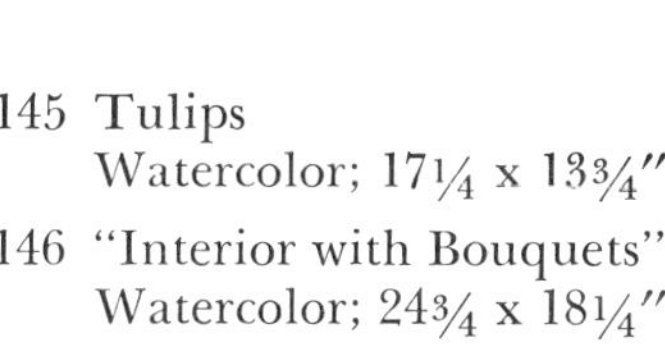

146

KNIGIN, Michael

Born   Brooklyn, New York, 9 December 1942.

Address   832-34 Broadway, New York, New York 10003.

Education   Tamarind Lithography Workshop, Los Angeles: 1964-65. Temple University, Tyler School of Art, Philadelphia: B.F.A., 1966.

Career   Artist. Assistant Professor of Graphics, Pratt Institute and Pratt Graphics Center, New York, 1974 to present. Formerly, costume designer, Metropolitan Opera, 1966. Worked at Collector's Press, 1966, and with Irwin Hollander, 1967. Instructor of graphics, Pratt Institute and Instructor of lithography, Pratt Graphics Center, 1968-73. Opened lithography workshop, New York, 1968. Co-owner and director, Chiron Press, 1968-72. Guest Lecturer, Herron School of Art, Indiana University, 1971. Instructor of Lithography, Yale University, 1971. Production and set designer, American Film Institute, Los Angeles, 1973. Technical director, The Burston Graphic Centre, Jerusalem, 1974.

Media   Lithography, painting.

One-person Exhibition   Upstairs Gallery, Southampton, Long Island.

Group Exhibitions   More than 40, including: Smithsonian Institution, Washington, D.C.; National Collection of Fine Arts, Washington, D.C.; Whitney Museum of American Art, New York; Associated American Artists, New York; American Pavilion, World Fair, Osaka, Japan, 1970; Mabat Gallery, Tel Aviv, Israel, 1975.

Awards   Ford Foundation Grant, 1964; E.A.T. Grant, MLA-ALA Lithographic Technical Institute, Local 1.

Collections   More than 20, including: the Whitney Museum of American Art, New York; Albright-Knox Museum, Buffalo; Museum of Graphic Art, New York; New York University Art Collection.

Works Published in   Knigin, M. and Zimiles, M. *The technique of fine art lithography.* New York, Van Nostrand Reinhold, 1969.
Print collector's newsletter, 1970.
Knigin, M. "Local choice: a note on the frontpiece," *Artist's proof,* Vol. XI, 1971.
Knigin, M. *Contemporary lithographic workshops around the world.* New York, Van Nostrand Reinhold, 1974.

147 "R.F.D."
Lithograph;
21½ x 30⅜"

KOYAMA, Tetsuo Michael

Born   Tokyo, Japan, 9 October 1933. Emigrated to the United States in 1964.

Address   ℅ New York Botanical Garden, Bronx, New York 10458.

Education   University of Tokyo: B.S., 1956; M.Sc., Plant Toxonomy and Plant Anatomy, 1958; Ph.D., Plant Taxonomy and Plant Anatomy, 1961.

Career   Curator, New York Botanical Garden, 1967 to present; (Adj.) Professor, City University of New York, 1971 to present. Botanist, specializing in taxonomic and anatomical research of phanerogams, especially Cyperaceae. Teaches botanical illustration, New York Botanical Garden. Formerly, Associate Professor, University of Ryukyus, 1958-59. Lecturer, Nippon University, 1960-61. Associate Professor, Tamagawa University, 1961-63. Educational Officer, Faculty of Science, University of Tokyo, 1961-63. Research Associate, New York Botanical Garden, 1963-64. Associate Curator, New York Botanical Garden, 1964-67. Professorship, Tamagawa University, 1976. Visiting Professor, Aarhus University, Denmark, 1977-78.

Media   Ink, watercolor, pencil.

Group Exhibitions   Hunt Institute Internationals, 1968, 1972.

Honors   Research Fellowship, National Research Council of Canada, 1961-63; United States Official Delegate, IX Pacific Science Congress, 1966; Medallion, University of Helsinki, 1968.

Collection   New York Botanical Garden.

Works Published in   (most recent publications)
Koyama, T. "New Cyperaceae from Nepal and adjoining Tibet," *Botanical magazine* (Tokyo), 1973.
Koyama, T. "The genus *Fimbristylis* (Cyperaceae)," *Ceylon botanical magazine* (Tokyo), 1974.
Koyama, T. "Smilacaceae, *"Flora of Thailand,* 1975.

148 *Arisaema thunbergii*
   Watercolor; 13⅞ x 8⅞"

149 *Arisaema heterophyllum*
   Watercolor; 13¾ x 9⅞"

Both lent by the artist

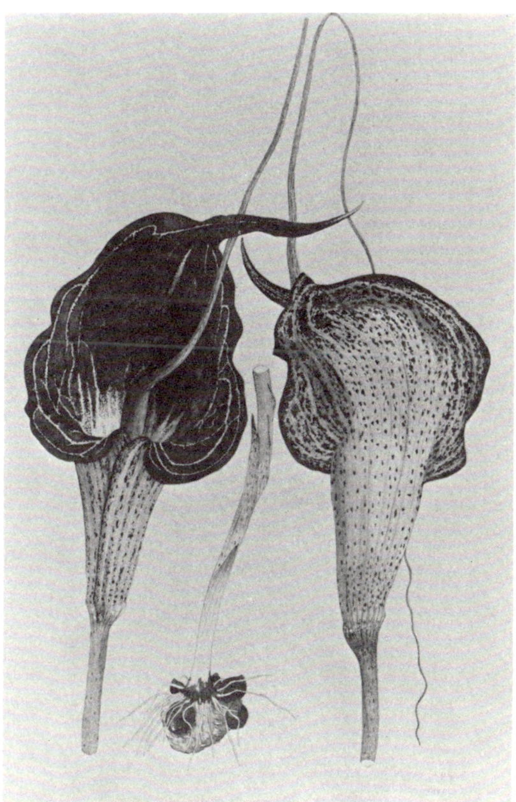

148

KUNZ, Otto Ludwig

Born Schwäbisch-Hall, Germany, 15 December 1904.

Address Am Sonnenweg 28, 7 Stuttgart 75, Heumaden, Germany.

Education Technischen Hochschule, Stuttgart: studied architecture; Diploma, Engineering, 1928; Ph.D., Engineering, 1937.

Career Retired free-lance artist and designer. Formerly, university lecturer, Hochschule, Stuttgart, and Professor of Fine Arts Drawing, Staatsbauschule, Stuttgart. Designer for various porcelain, paper, and fabric industries.

Medium Watercolor.

One-person Exhibitions Kongresshaus, Baden-Baden, 1974; Städt. Galerie Fauler Pelz, Ueberlingen a.B., 1974; Galerie Maercklin, Stuttgart, 1976; Kunsthaus Schaller, Stuttgart, 1977.

Works Published in *Blumenkalender*. Stuttgart, Verlag Staehle und Friedl, 1949 to present.*Rosenkalender* and *Meister der Flora* (calendar) by the same publisher.

150 Oriental Poppy, *Papaver orientale**
Watercolor; 17½ x 13½"

151 *Iris sibirica;* Columbine,*
Aquilegia sp.; Pansy, *Viola tricolor;*
Bleeding Heart, *Dicentra spectabilis*
Watercolor; 13½ x 11⅛"
Gift of the artist

150

LANGEDIJK, Gerrit Jan

Born   Wageningen, Netherlands,
9 July 1930.

Address   Swammerdamlaan 26,
Bennekom, Netherlands.

Education   Academy of Art,
Arnhem: painting and graphic
art, 1951.

Career   Sculptor. Formerly, part-
time botanical artist, State
Agricultural University,
Wageningen, 1961-73.

Media   Sculpture, watercolor, oil, ink.

Group Exhibitions   1,000 Years of Flower Illustration, the
National Meerman and Westreenen, Museum, The Hague,
1966; Stockholm, 1967; Galerie "De Beerenburght,"
Eckenweil, Netherlands, 1976.

Collection   State Agricultural University, Wageningen.

Works Published in   *Miscellaneous papers VI.* Wageningen,
Landbouwhogeschool, 1970.
Nannenga, N.E. *Notes on Hedera species, varieties and
cultivars grown in the Netherlands.* Bremekamp, 1970.
Breteler, F.J. *The African Dichapetalaceae.* Wageningen,
H. Veenman & Zonen NV, 1973-74.*
Jong, P.C. de. *Flowering and sex expression in Acer L.*
Wageningen, 1976.

152 *Dichapetalum angolense**
Ink; 14⅝ x 9⅛"

153 Apple, *Malus* 'Van Eseltine'
Ink; 11⅞ x 8¼"

Permanent loans of the Laboratorium voor
Plantensystematiek en -Geografie,
Landbouwhogeschool, Wageningen, Netherlands

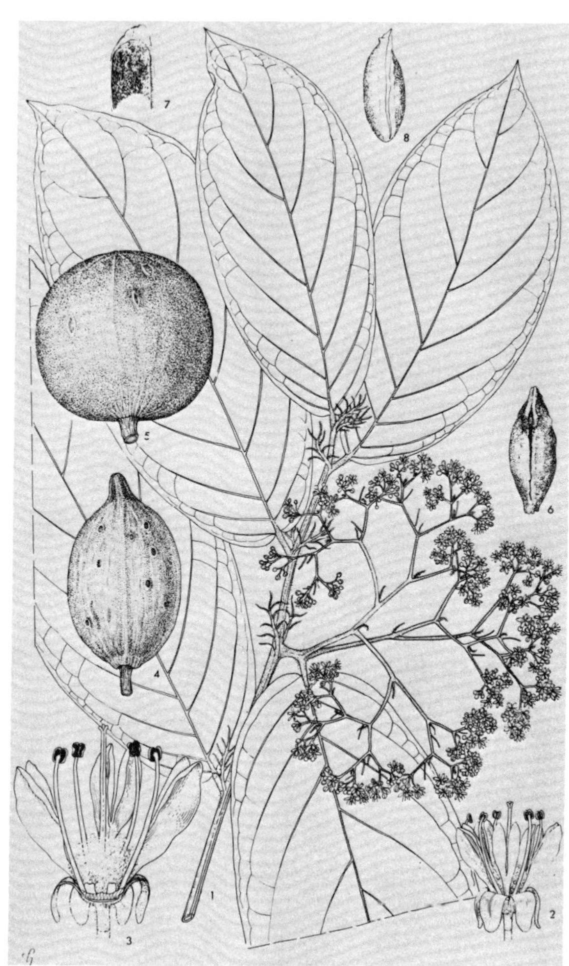

152

LEATHERWOOD, Linda Ann Baumhardt (Mrs. Donald H.)

Born   Evanston, Illinois,
27 November 1948.

Address   24337 Gosling Road,
Spring, Texas 77379.

Education   Ringling School of Art,
Sarasota, Florida: 1966-68; studied
drawing under Frederick Sweeney.
University of South Florida,
Tampa: Botany, 1969-71; began
ink illustration under Dr. Robert
W. Long.

Career   Free-lance illustrator and oil landscapist. Formerly,
Botanical Illustrator, University of South Florida Herbarium,
1970-71. Commercial artist, and assistant at the University
of South Florida Botanical Gardens. Took on private
commissions from 1974 to present.

Media   Ink, oil.

One-person Exhibitions   University of South Florida,
permanent display; Herbarium, University of South Florida,
1971-77; Punta Gorda Library, Punta Gorda, Florida, 1972.

Group Exhibitions   Gasparilla Art Festival, Tampa, 1970;
Northwest Fine Arts League, Houston.

Award   Best of Media, Best of Show (oil landscape),
Northwest Fine Arts League, 1977.

Collections   University of South Florida Herbarium;
Dr. Clinton J. Dawes.

Commissions/Works Published in   Long, R.W. and Lakela, O.
A flora of tropical Florida. Coral Gables, Florida,
University of Miami Press, 1971. *

Phycologia, 11 (3/4), 1972.

Lakela, O. and Long, R.W. *Plants of the Tampa Bay area*
(cover illustration), 1970.

Dawes, C. *Marine algae of the west coast of Florida.* Coral
Gables, University of Miami Press, 1974.

Lakela, O. and Long, R.W. *Ferns of Florida.* Miami,
Banyan Books, 1976.

Fleming, G., Genelle, P. and Long, R.W. *Wild flowers of
Florida.* Miami, Banyan Books, 1976 (title-page
illustration).

Dawes, C. *Flora of marine algae of Florida* [in progress].

154 *Scaevola plumieri* *
Ink; 12⅝ x 10″

155 Yellow Pond-Lily,
Nuphar luteum spp. *macrophyllum* *
Ink; 13¼″ x 10¾″

156 *Eyrngium aromaticum* *
Ink; 14 x 10⅝″

Gifts of the artist

156

LEXA-REGÉCZI, Marta

Born   Budapest, Hungary, 4 March 1918.

Address   Orbánhegyi - út 51, 1126 Budapest, Hungary.

Education   Teacher's Academy, Convent of the Order of I.B.M.V.: diploma, 1937. Studied at Jaschik's Private School of Graphic Arts.

Career   Botanical illustrator, Academie Horti et Viticulture, Budapest, 1963 to present. Formerly, free-lance botanical illustrator, specializing in mushrooms.

Media   Watercolor, ink.

Group Exhibitions   Hunt Institute Internationals, 1968, 1972; International Exhibition of Botanical Art, Johannesburg, 1973; North American Mycological Association Art Salon, Dartmouth, New Hampshire, 1975.

Collection   Academie Horti et Viticulture, Budapest.

Works Published in   *Instructions for phenological observations.* 1967.
General botany. Vols. 1-2, 1967.
Meteorological development of plants in agriculture. 1967.
Kárpáti, Z. *Kertészeti Növénytan (Horticultural botany).* Budapest, Mezögazdasági Könyvkiadö, 1968.
Élövilág (Living world). 1968, 1972.
Kárpáti, Z. *A Növények Világa (The life of plants).* Vol. II, 1969.
Kerekes, J. *Cyógynövénytermesztés (Cultivation of herbs).* Budapest, Mezögazdasági Könyvkiadö, 1969.
Biológiai Lexikon (Encyclopedia of biology). 1972.
Debreczy. *Atlas of dendrology.* Akadémia Kiadó [in progress].
Kádas, Lajos. *Book of fruits.* Budapest, Móra Kiadó [in progress].
Mezógazdasági Lexikon (Encyclopedia of agriculture) [in progress].
Various Hungarian horticultural periodicals.

157   Raspberry, *Rubus* hybrid
Watercolor; 7½ x 7¼″

158   Cowslip, *Primula veris*
Watercolor; 10⅛ x 7″

158

LID, Dagny Tande (Mrs. Johannes)

Born   Nissedal, Norway, 25 May 1903.

Address   Minister Ditleffs vei 20, Oslo 8, Norway.

Education Kongsgård, Stavanger: Certificate of Examen Artium, 1924. Statens Håndverks- og Kunstindustriskole, Oslo: 1926-27. Statens Kunstakademi, Oslo: 1928-29.

Career   Botanical artist and illustrator.

Media   Watercolor, ink.

One-person Exhibitions   Kunstnerforbundet, Oslo, 1965; Kunstforeningen, Trondheim, 1974; Porsgrunn senteret, Oslo, 1974; Kunstforeningen, Porsgrunn, 1976; Kunstforeningen, Larvik, 1977.

Group Exhibitions   Hunt Institute International Exhibition, 1972; Hunt Institute Travel Shows (International).

Works Published in   Schøyen & Jørstad. *Diseases in gardens.* 1942.
Christopherson, E. *Medicinal plants.* 1943.
Lid, J. *Norsk flora.* 1944 (rev. ed. 1952).
Löve, A. *Islenzkar jurtir.* 1945.
Norwegian potatoes. 1951.
Gjaerevoll, O. and Jørgensen, R. *Fjellflora.* Swedish and Finnish editions, 1952. *Mountain flowers of Scandinavia* (English ed.). Trondheim, 1963.
Porsild, A.E. *Edible plants of the Arctic.* 1953.
Porsild, A.E. *Illustrated flora of the Canadian Arctic.* 1957.
Polunin, N. *Circumpolar flora.* 1959.
Lid, J. *Norsk og Svensk flora.* 1963 (to be reprinted, 1978).
Lid, J. *The flora of Jan Mayen.* 1964.
Ronning, O. *Svalbards flora.* 1964.
Lid, J. *Contributions to the Canary flora.* 1966.
Hultén, E. *Manual of the Alaska flora.* 1966.
Lid, J. *Blomsterboka.* Oslo, Det Norske Samlaget, 1966 (rev. ed. 1968).
Löve, A. *Islenzk ferdaflora.* 1970.
Lid, D.T. *Å nei for en vår.* Oslo, H. Aschehoug & Co. (W. Nygaard), 1972.
Lid, D.T. *Lykken mellon to mennesker.* Oslo, H. Aschehoug & Co. (W. Nygaard), 1974.
Lid, D.T. *Syng blomstring . . . .* Oslo, H. Aschehoug & Co. (W. Nygaard), 1975.
Porsild, A.E. *Rocky mountain wildflowers.* Ottawa, National Museum of Canada, 1975.
Lid, D.T. *Vindens lek.* Oslo, H. Aschehoug & Co. (W. Nygaard), 1977.
Designs for porcelain coffee service, Porsgrunds Porselaensfabrik, Porsgrunn, 1972. *

159 Designs for stamps: Violet, *Viola biflora;* Rock Speedwell, *Veronica fruticans; Phyllodoce caerulea.* (Shown with matching stamps) Watercolor and ink; 4 x 3¼″ average size Gifts of the artist

160 Mountain Flora Coffee Set (cups, saucers, plates) produced by Porsgrunds Porselaensfabrik, Norway. Alpine Cinquefoil, *Potentilla crantzii;* Rock Speedwell, *Veronica fruticans;* Purple Saxifrage, *Saxifraga oppositifolia.* *
Felspar porcelain; 6⅝″ plate diameter
Lent by Dr. Gilbert S. Daniels

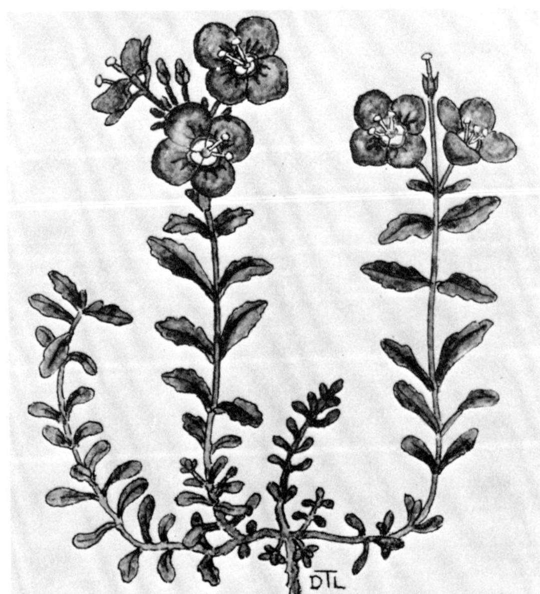

159

LOEWER, Henry Peter

Born   Buffalo, New York, 13 February 1934.

Address   Cochecton Center, New York 12727.

Education   University of Buffalo, Albright Art School: B.F.A., Graphics and Art History, 1958.

Career   Director, Graphos Studio, Inc., New York (scientific and book illustration), 1970 to the present. Formerly, United States Army Cryptographer, 1959-61. Self-employed illustrator, 1961-63. President, Loewer Studio, Inc., New York (scientific and book illustration). Numerous civic positions between 1971-77, including Secretary of Cochecton Planning Board, and Councilman, Cochecton Town Board.

Media   Pencil, ink, watercolor, etching, engraving.

One-person Exhibition   Catskill Art Center, 1977.

Group Exhibitions   Catskill Art Center, 1972, 1974, 1976, 1977; Hunterdon Art Center, Clinton, New Jersey, 1977.

Collections   Many private collections.

Works Published in   Over 300 books published by the Macmillan Company (College Department).

Bayer and Owre. *Free-living lower invertebrates.* New York, Macmillan, 1968.

Montague, A. *Man: his first two-million years.* New York, Columbia University Press, 1969.

Spear, A.T. *Brancusi's birds.* New York, New York University Press, 1969.

Simmons, A.E. *Growing unusual fruit.* New York, Walker, 1972.

Goodman and Gilman. *Pharmacological basis of therapeutics.* New York, Macmillan, 1975.

Miles, B. *Wildflower perennials for your garden.* New York, Hawthorne, 1976.*

Loewer, H.P. *Indoor water gardener's how-to handbook.* New York, Walker, 1973.

Loewer, H.P. *Bringing the outdoors in.* New York, Walker, 1974.**

Loewer, H.P. *Seeds and cuttings.* New York, Walker, 1975.

Window gardening (ed.). New York, 1977.

Loewer, H.P. *Growing and decorating with grasses.* New York, Walker, 1977.

Loewer, H.P. "The Catskill gardener," Sullivan County *Democrat* (weekly column).

Illustrations for *Natural history, Search, Garden, Green scene, House & garden, Woman's day.*

161   Queen's Tears, *Billbergia nutans***
Ink; 10¼ x 8¼″
Gift of the artist

162   Moon Vine, *Ipomoea noctiflora***
Ink; 12½ x 8½″
Gift of the artist

163   Cobra Lily, *Darlingtonia californica* (left);
Trumpet Pitcher Plant, *Sarracenia flava* (right)**
Ink; 12¼ x 9½″
Gift of the artist

164   Terrarium **
Ink; 13½ x 9½″
Gift of the artist

165   Camas-Lily, *Camassia cusickii* *
Ink; 10 x 6¼″

166   Canadian Burnet, *Sanguisorba canadensis**
Ink; 10 x 6¼″

161

LONG, Lois

Born   Clarksdale, Mississippi, 13 August 1918.

Address   8 West 13th Street, New York, New York 10011.

Education   Pratt Institute, New York: 1936-39. Art Students League, New York: 1946-47.

Career   Free-lance graphic artist.

Media   Pencil, watercolor, tempera, lithography.

One-person Exhibitions   Carl Solway Gallery, Cincinnati, 1973; Cranbrook Art School, Bloomfield Hills, Michigan.

Group Exhibitions   New York City: Leo Castelli Gallery; Martha Jackson Gallery; Museum of Modern Art; New York Public Library; Rizzoli Gallery.

Collections   Betty Parsons Gallery; University of California, Santa Cruz; Museum of Modern Art; Yale University; Stendenlijk, Amsterdam; Tokyo Gallery, Japan.

Works Published in   *Natural history magazine; New York times; Book of the Month Club news; Cranbrook alumni magazine.*
Long, L., Cage, J., and Smith, A. *Mushroom book.* New York, Hollanders Workshop Inc., 1972.*

167 Velvet-stemmed Collybia Mushroom,
   *Collybia velutipes**
   Color lithograph;
   21½ x 15″ sheet size

168 Morel Mushroom, *Morchella* sp.*
   Color lithograph;
   21½ x 15″ sheet size

169 Chicory, *Cichorium intybus*
   Pencil and watercolor; 14¼ x 21″

167

LUCIONI, Luigi

Born   Malnate, Italy, 4 November 1900; has lived in the United States since 1911.

Address   33 West 10th Street, New York, New York 10011.

Education   Cooper Union, New York, 1916-20. National Academy of Design: studied with William Auerbach Levy, 1920-26.

Career   Painter and printmaker. Formerly, teacher, Art Students League, New York, 1932.

Media   Oil, watercolor, etching, engraving.

One-person Exhibitions   Association of American Artists, 1952, 1976; Shelburne Museum, Shelburne, Vermont, 1968; "The Art of Luigi Lucioni" (film), Vermont Educational Television Station, 1969-72; several shows at the Ferargil Galleries, New York.

Group Exhibitions   Numerous, including: Carnegie Institute International Exhibitions; Corcoran Gallery of Art, Biennial Exhibition; Hunt Institute International, 1972; Hunt Institute Travel Shows (Prints).

Honors/Awards   First Popular Prize, International Exhibition, Carnegie Institute, 1939; First Popular Prize, Biennial Exhibition, Corcoran Gallery, Washington, D.C.; Prize, National Academy of Design; Tiffany Medal.

Collections   Numerous public collections, including: the Metropolitan Museum of Art, New York; Whitney Museum of Art, New York; Fogg Art Museum, Boston; Addison Gallery of American Art, Andover, Massachusetts; Library of Congress, Washington, D.C.; Carnegie Institute, Pittsburgh; Victoria and Albert Museum, London; Shelburne Museum, Shelburne, Vermont; Middlebury College Museum, Connecticut; New York Public Library; Pennsylvania Academy of Design, Philadelphia.

Works Published in   *Life, Town and country, Vermont life, American artist.*

170 "Between the Birches," *Betula* sp.
  Etching; 9 x 13$\frac{1}{16}$" plate mark

171 "Spreading Maple," *Acer* sp.
  Etching; 10$\frac{7}{8}$ x 15$\frac{1}{2}$" plate mark

170

MACKAY, Donald A.

Born   Halifax, Nova Scotia, 13 August 1914.

Address   12 Glendale Road, Ossining, New York 10562.

Education   Massachusetts School of Art, Boston. Pratt Institute Center for Printmaking, New York. University of Morelia, Mexico: studied graphics with Alfredo Zalce.

Career   Printmaker and free-lance artist.

Media   Etching, lithography, watercolor, drawing.

One-person Exhibitions   Vanderbilt University, Nashville; Briarcliff College, Briarcliff Manor, New York.

Group Exhibitions   American Watercolor Society; Society of American Graphic Artists; Boston Print Club; Philadelphia Print Club; Silvermine Guild; Pratt Institute International Print Show; Smithsonian Institution traveling exhibition (50 American Drawings); Hudson River Museum International Exhibition, New York.

Collections   Minnesota Museum of Art; Vanderbilt University.

Commissions/Works Published in   *New York times, Newsweek, Graphis, Esquire, Time, Life, National geographic,* and other periodicals.
Children's books for Viking, Bradbury Press, James H. Heineman, and Dial Press.

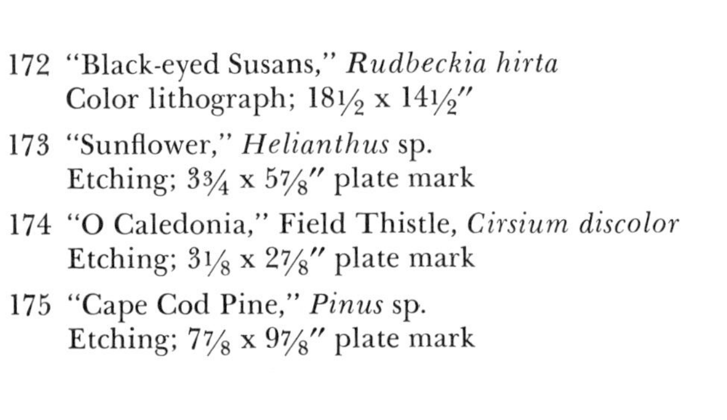

172   "Black-eyed Susans," *Rudbeckia hirta*
Color lithograph; 18½ x 14½"

173   "Sunflower," *Helianthus* sp.
Etching; 3¾ x 5⅞" plate mark

174   "O Caledonia," Field Thistle, *Cirsium discolor*
Etching; 3⅛ x 2⅞" plate mark

175   "Cape Cod Pine," *Pinus* sp.
Etching; 7⅞ x 9⅞" plate mark

172

MAGGIORA, Laura Rosano

Born   Turin, Italy, 3 April 1938.

Address   Strada del Lauro 43, 10132 Turino, Italy.

Education   Albertina Academy, Turin: courses in decoration, 1961. Studied with Professor Cerutti, National Sciences Faculty, Turin.

Career   Medical and scientific illustrator. Illustrator of children's fiction.

Media   Watercolor, acrylic.

Group Exhibition   8th Exhibition of Illustrators, Bologna, 1974.

Works Published in   von Frieden, L. *I funghi de tutti i paesi,* Milan, Rizzoli, 1964.
Bresadola, G. *Funghi mangerecci e velenosi.* Trento, Monauni, 1965.
Rinaldi, A. *L'atlante dei funghi.* Milan, Mondadori, 1972.*
Illustrations for science textbooks published by Paravia, Turin.

176 Mushrooms, *Amanita flavoconia; A. formosa; A. shiaolita; A. regalis; A. aureola**
Watercolor; 9¾ x 9½"

177 Mushroom, *Gyromitra esculenta* *
Watercolor; 7¾ x 8⅛"
Gifts of Mondadori Publishers, Milan, Italy

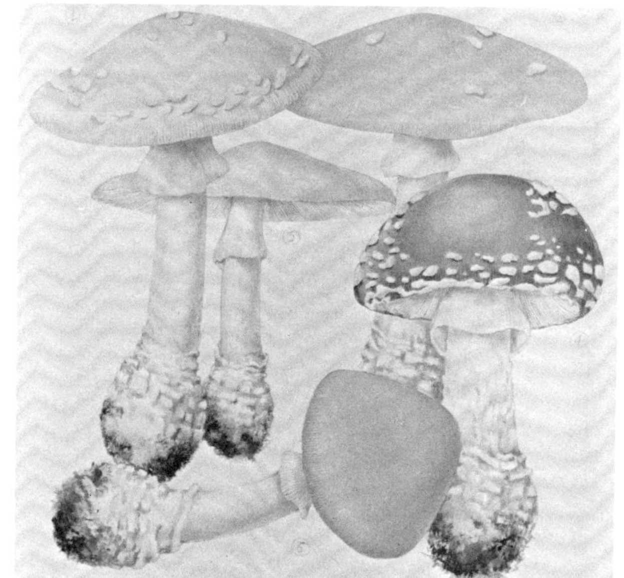

176

MALTZMAN, Stanley

Born   New York, New York,
4 July 1921.

Address   Rt. 1, Box 158, Freehold,
New York 12431.

Education   Phoenix School of
Design, New York: studied
commercial art, 1946-48.
Self-taught fine artist.

Career   Printmaker and painter.
Formerly, art director, Vick
Chemical Company.

Media   Charcoal, lithography, etching, watercolor, casein.

One-person Exhibitions   Hudson River Museum, 1964; Frank
Rehn Gallery, 1974; Hartley Gallery, 1975; Green County
Council on the Arts Gallery, 1977.

Awards   Childe Hassam Fund Purchase Award, American
Academy of Arts and Letters; Gold Medal, Graphics, Hudson
Valley Art Association; Gold Medal, Graphics, American
Artists Professional League; Gold Medal, Graphics, American
Veterans Society of Artists; Ball State University Award;
Gold Medal, Graphics, Academic Artists Association; First
Prize, Graphics, Connecticut Academy of Fine Arts.

Collections   Eisenhower College, Salem, New York;
Pleasantville Middle School, Pleasantville, New York; Hudson
River Museum, New York; A.T.&T. Company, Chicago
Division.

Group Exhibitions   State University College, New Paltz, New
York; National Arts Club, New York; Ball State University,
Muncie, Indiana; Institute of Man and Science,
Rensselaerville, New York; Norfolk Museum of Arts and
Sciences, Virginia; Hudson River Museum, New York;
Springfield Museum, Massachusetts; Hammond Museum,
North Salem, New York; Smithsonian Institution traveling
exhibition; National Academy/American Academy of Arts
and Letters; Albany Institute of History and Art, New York.

Commissions/Works Published in   Martin, B., Jr. and
Brogan, P. *Sounds of a distant drum* and *Sounds of mystery*.
New York, Holt, Rinehart & Winston, 1972.
Commissions from Tiffany Company and Steuben Glass
Company, New York.
Steuben catalogue, spring 1976.

178   "Winter Cats," *Typha* sp.
Lithograph; 23 1/16 x 18"

179   "Thistle," *Cirsium* sp.
Lithograph; 8 1/4 x 12"

178

McCUBBIN, Charles Wilson

Born   Melbourne, Victoria, Australia, 24 August 1930.

Address   6 Manniche Avenue, Box Hill North, Victoria 3129, Australia.

Education   Royal Melbourne Institute of Technology: 1949, 1965-67. Studied portrait painting with Murray Griffen.

Career   Free-lance artist and illustrator.

Media   Watercolor, Oil.

One-person Exhibitions   Joshua McClelland Print Room, 1963, 1965, 1968; South Yarra Galleries.

Collection   Commonwealth Art Advisory Council, Canberra.

Works Published in   McCubbin C. *Australian butterflies*. Melbourne, Thomas Nelson (Australia) Ltd., 1971.

180 "Phoenix," Australian native plants regenerating after fire
Watercolor and gouache; 10½ x 7¼"

McEWEN, Roderick (Rory)

Born   Polwarth, Scotland,
12 March 1932.

Address   9 Tregunter Road,
London S.W. 10, England.

Education   Cambridge University:
B.A., Literature, 1955.

Career   Free-lance artist. Formerly,
Art Editor of *The spectator*,
London, 1956-60.

Media   Watercolor on vellum,
etching.

One-person Exhibitions   Durlacher Brothers Gallery, New
York, 1962, 1964; André Weil, Paris, 1964; Douglas & Foulis,
Edinburgh, 1966; Byron Gallery, New York, 1967; Richard
Demarco, Edinburgh, 1968; Redfern Gallery, London, 1972,
1974, 1976; Sonnabend Gallery, New York, 1973.

Group Exhibitions   The National Library, Edinburgh, 1964;
Hunt Institute International, 1964; Scottish Arts Council,
1976.

Collections   The White House, Washington, D.C.; National
Gallery of Modern Art, Edinburgh; Victoria & Albert
Museum, London; many private collections.

Works Published in   Moreton, C.O. *Old carnations and pinks.*
London, George Rainbird, 1955.
Moreton, C.O. *The auricula.* London, The Ariel Press
Limited, 1964.
Blunt, W. *Tulips & tulipomania.* London, Basilisk Press, 1977.*
Etchings published by Ganymed Original Editions Ltd.,
London.

181   Old English Florist Tulip,
Tulipa 'Sam Barlow'*
Watercolor on vellum;
23 x 20″ sheet size
Lent by the artist

182   Old English Florist Tulip,
Tulipa 'Bessie' *
Watercolor on vellum;
23 x 20″ sheet size
Lent by the artist

183   Onion, *Allium* sp.
Etching; 10½ x 7″ plate mark

184   Pomegranate, *Punica granatum*
Etching; 13¼ x 16⅝″ plate mark

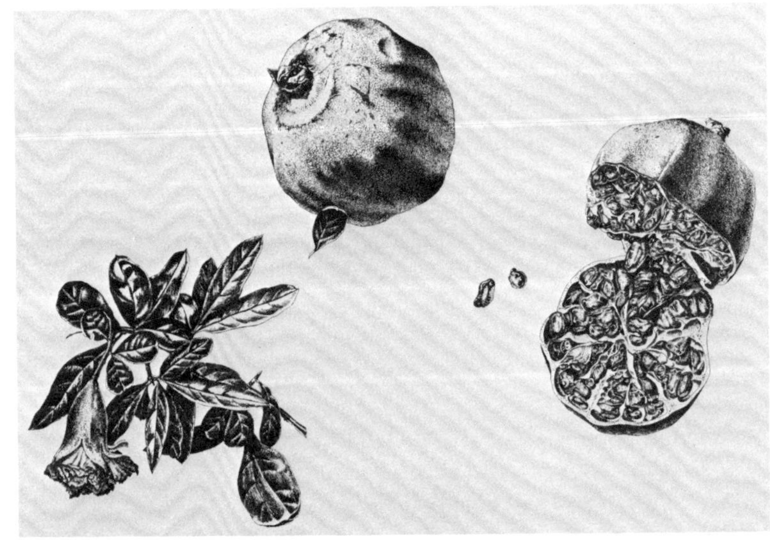

184

MOCKEL, Henry R.

Born   Berlin, Germany, 14 May 1905. Has lived in the United States since 1923.

Address   P.O. Box 725, Twentynine Palms, California 92277.

Education   Gymnasium, Berlin, Germany: 1911-23. City College of New York: studied engineering, 1927-30. Grand Central School of Art: 1931-35.

Career   Illustrator, printmaker, and co-owner (with Beverly Mockel) of the Pioneer Art Gallery, Twentynine Palms. Formerly, worked for New York Telephone Company for 15 years. Private art teacher, Bronx School of Art, for 3 years. Artist and farmer in Maine, and in Cooperstown, New York, 1946-57.

Media   Serigraphy, oil, acrylic, watercolor, scratchboard, woodcut, etching, pre-separated color illustrations.

One-person Exhibitions   Numerous, including: Riverside Museum, Riverside, California, 1969; Desert Art League, China Lake, California, 1969; Desert Reserve, Palm Desert, California, 1975, 1977; San Bernardino County Library, Twentynine Palms, 1976; University of California, Riverside, 1976; Edward Dean Museum, Cherry Valley, California, 1977; Oregon State University, Corvallis, 1977.

Group Exhibitions   Numerous, including: Los Angeles County Museum of Natural History, 1968; Hunt Institute Internationals, 1964, 1968; Southern Vermont Art Center, Manchester, 1965, 1976; XI International Botanical Congress, Seattle, Washington, 1969; Universidad National, Buenos Aires, 1972; International Exhibition of Botanical Art, Johannesburg, 1973; University of Durban, University of Witwatersrand, and City Hall in Pietermaritzburg, South Africa, 1973; Moorten's Desert Botanic Garden, Palm Springs, California, 1976; Palm Springs Art Museum, California, 1977.

Awards   Second Prize, Gewerbe Austellung, Berlin, 1922; First Prize Medal, Grand Central School of Art, New York, 1933; Presentation Print, Print Club of Albany, New York, 1964; John Taylor Arms Memorial Purchase Award, Albany, New York, 1965; Presentation Print, Rochester Print Club, New York, 1969.

Collections   Over twenty institutions, including: California State Library, Sacramento; Department of the Interior; Richfield Oil Corporation, Los Angeles; Santa Barbara Botanic Garden; United States National Arboretum, Washington, D.C. Many private collections.

Works Published in   *Richfield wildflower book.* Los Angeles, 1966.
Mockel, H.R. and B. *Hot air from the desert.* Twentynine Palms, Henry and Beverly Mockel, 1968.
Mockel, H.R. and B. *Mockel's desert flower notebook.* Twentynine Palms, Henry and Beverly Mockel, 1971.
Raven, P. *The genus Camissonia.* Washington, D.C., Smithsonian Institution.
Cactus and succulent journal, 1967; *Desert magazine,* 1961-65; *California horticultural journal,* October, 1972.
16 note cards for Los Angeles State and County Arboretum.

185 California Bluebell,
Phacelia campanularia
Acrylic; 21½ x 27½"

MONSMA, Mary Susan

Born   Washington, D.C., 26 May 1951.

Address   304 Irwin Street, Silver Spring, Maryland 20901.

Education   Calvin College, Grand Rapids, Michigan: B.A., 1974.

Career   Free-lance illustrator, Department of Botany, Smithsonian Institution. Formerly, part-time elementary school art teacher, 1974-76.

Medium   Ink.

Group Exhibition   Guild of Natural Science Illustrators, 1976.

Collection   Smithsonian Institution.

Works Published in   "Argentine *Solanum*," *Kurtziana*, Department of Botany, University of Cordoba. *Ferns of Costa Rica and the Choco* [in progress].

186 *Pleopeltis* spp.
  Ink on acetate; 16½ x 10½"

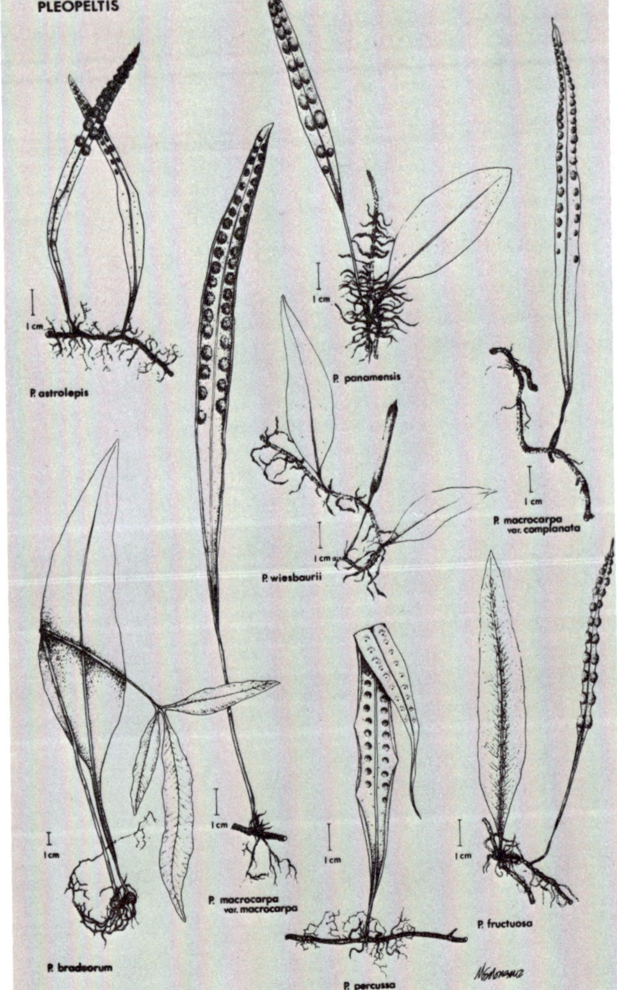

MORAN, Margaret Ann

Born   Cleveland, Ohio, 11 January 1942.

Address   5635 South Dorchester, Chicago, Illinois 60637.

Education   Cleveland Institute of Art: 1956-60. Art Institute of Chicago: B.F.A., 1967.

Career   Director, Snowgoose Graphics.

Media   Ink, wash, color pencil, watercolor, acrylic, mixed media.

Group Exhibition   Chicago Creativity, 1977.

Collection   Institute for Arctic and Alpine Research, Boulder, Colorado.

Works Published in   Solem, G.A. *Endodontoid land snails from Pacific islands.* Chicago, Field Museum of Natural History, 1976.
The Encyclopaedia britannica.
The World book encyclopedia.
The World book dictionary.
Science year.
Mariah magazine.
Illustrations for Ethical Communications Inc.; Houghton Mifflin Company; Silver Burdett and Company; Harper and Row, Publishers; Charles E. Merrill Publishing Company.
Slide programs for the United States Forest Service.

187   "Two Buds," Bindweed, *Convolvulus* sp.
Hand-colored etching; 10¾ x 7⅜″ plate mark

188   Japanese Maple, *Acer palmatum* (top);
Silver Maple, *Acer saccharinum* (center);
Box Elder, *Acer negundo* (bottom)
Ink; 11¼ x 4½″

189   Woody dicot stem (cross-section)
Ink; 10½ x 8″

190   *Crocus* sp.
Ink; 6¼ x 5½″

All lent by the artist

187

MOSER, Arthur Barry

Born   Chattanooga, Tennessee, 15 October 1940.

Address   Pennyroyal Press, One Railroad Street, Easthampton, Massachusetts 01027.

Education   Auburn University, Auburn, Alabama: Industrial Design, 1958-60. University of Chattanooga, Tennessee: B.S., Painting, 1960-62. Studied painting, drawing, and printmaking with George Cress, Leonard Baskin, Jack Coughlin, and Harold McGrath.

Career   Printmaker. Instructor of Graphic Arts and Art History, Williston-Northampton School, Easthampton, Massachusetts. Private pressman, Pennyroyal Press, Easthampton. President, Hampshire Typothetae, Northampton. Visiting artist and lecturer at many schools and other institutions.

Media   Ink, etching, wood engraving, letterpress.

One-person Exhibitions   At numerous institutions, including: Hunter Gallery, Chattanooga, 1964; Graphic Arts Gallery, Springfield, Massachusetts, 1969; Unitarian Society, Amherst, Massachusetts, 1969; University of Tennessee, Chattanooga, 1970, 1975; Springfield Public Library, Springfield, Massachusetts, 1973; Berkshire Museum, Pittsfield, Massachusetts, 1973; Rhode Island College, Providence, Rhode Island, 1975; Smith College, Northampton, Massachusetts, 1976; Boston Athenaeum, Boston, 1976; University of Nebraska, Omaha, 1976; Holyoke Museum Wistariahurst, Holyoke, Massachusetts, 1977; New Hampshire Technical Institute, Concord, 1977.

Group Exhibitions   More than 30, including: Library of Congress, Washington, D.C., 1975; Krakow, Poland, 1975; University of Wisconsin, Milwaukee, 1976; Houghton Library, Harvard University, Cambridge, Massachusetts, 1976.

Honors/Awards   Numerous, including: elected member, XIth International Botanical Congress, 1969; Second Prize, Cape Cod Annual, 1971; Award of Merit, New Hampshire International, 1974.

Collections   More than 30, including: Universities of Massachusetts, Tennessee, California, and North Carolina; Harvard University, Princeton University; New York Botanical Garden; Missouri Botanical Garden, Morton Arboretum; numerous private collections.

Works Published in   Numerous publications containing plant illustrations, including:
The death of the Narcissus. Castalia Press, 1970.
Bacchanalia. Easthampton, Pennyroyal Press, 1970.
Amahjian-Ahles. *The flora of Massachusetts.* Amherst, University of Massachusetts Press.
Smyth. *Thistles and thorns.* Omaha, University of Nebraska, Abattoir Editions.
Ramsey, P. *Eve singing.* Easthampton, Pennyroyal Press, 1977.
Falk. *Song of songs.* New York, Harcourt, Brace, Jovanovich, 1977.
Numerous other works published by:
Yankee magazine; Publishers weekly; The New York review of books; Pennyroyal Press; The University of Massachusetts Press; The University Press of New England; The University of Texas; Houghton-Mifflin Company, Boston; Little, Brown Co., Boston; Harcourt, Brace, Jovanovich; The Brown University Press.

191 "Garden of Satanic Delights"
Wood engraving; 7⅞ x 13″

192 Jack-in-the-Pulpit,
Arisaema triphyllum
Etching; 7 x 5″ plate mark

193 Trailing Arbutus, *Epigaea repens*
Etching; 6¾ x 4⅞″ plate mark

194 Purple Trillium, Stinking Benjamin,
Trillium erectum
Etching; 6⅞ x 5¼″ plate mark

195 Pink Lady's-Slipper,
Moccasin Flower,
Cypripedium acaule
Etching; 7⅜ x 4⅞″ plate mark

191

NANAO, Kenjilo

Born   Aomori, Japan, 26 July 1929. Has lived in the United States since 1960.

Address   640 Santa Rosa Avenue, Berkeley, California 94707.

Education   Nikon University. Asagaya Art Institute. Tamarind Lithography Workshop, Los Angeles: 1968-69. Brooklyn Museum Art School. San Francisco Art Institute: M.F.A., 1970.

Career   Printmaker and Associate Professor of Art, California State University, Hayward, 1971 to present. Formerly, printer at Collector's Press, San Francisco, 1970. Printmaking instructor, California State University, Hayward and San José, 1970.

Media   Lithography, drawing.

One-person Exhibitions   American Institute of Physics, New York, 1963; Brooklyn Museum Art School, New York, 1964; Tsubaki Kindai Gallery, Tokyo, 1965; Upstairs Gallery, Cincinnati, 1969; Galerie Smith-Andersen, Palo Alto, 1971, 1974; Kanecho Gallery, Aomori, 1971; The Santa Barbara Museum of Art, California, 1972; Hartnell Junior College, Salinas, California, 1973; Achenbach Foundation for Graphic Arts, Palace of the Legion of Honor, San Francisco, 1973; Richmond Art Center, California, 1977.

Group Exhibitions   More than 40 since 1963, including: Fulton Gallery, New York, 1963; San Francisco Hall of Justice Exhibition, 1967; San Francisco Museum of Art, 1967, 1973; Palace of the Legion of Honor, San Francisco, 1971; Brooklyn Museum of Art, 1973; Cincinnati Art Museum, 1973; Honolulu Academy of Art, 1973, 1975; Library of Congress, Washington, D.C., 1973, 1975; Associated American Artists, New York, 1974; Hunt Institute, 1975; National Museum of Modern Art, Tokyo, 1976.

Awards   Ford Foundation Grant, Tamarind Lithography Workshop, Los Angeles, 1968-69; Award from KQED Television, San Francisco; Selected by the San Francisco Art Commission for Exchange Show to Vancouver, British Columbia, 1972. Purchase Prizes: San Francisco Art Festival, 1970-72; Brooklyn Museum of Art, 1972; Honolulu Academy of Arts, 1973; City of Philadelphia.

Collections   Cincinnati Museum of Art; Honolulu Academy of Art; Achenbach Foundation for Graphics, Palace of the Legion of Honor, San Francisco; Brooklyn Museum of Art; Home Savings and Loan Corporation, Los Angeles; Museum of Modern Art, New York; Pasadena Art Museum; Los Angeles County Museum; Grunewald Collection, University of California, Westwood; Rochester Institute of Technology; Minnesota Museum of Art, St. Paul; Library of Philadelphia; The Oakland Museum; Library of Congress.

Commissions   Editions of prints for: New Leaf Publisher, Walnut Creek, California; Associated American Artists, New York; City of San Francisco.

196 "Day Plant," Cactus
Color lithograph;
20½ x 26″ sheet size

NICHOLSON, Barbara Evelyn

Born   Surrey, England, 15 November 1906.

Address   Carpenters, Donhead St. Mary, Shaftesbury, Dorset SP7 9DG, England.

Education   Southend-on-Sea Municipal Art School: 1924-26. Royal College of Art, Design School, South Kensington, London: Associateship, 1929.

Career   Free-lance botanical illustrator. Formerly, landscape oil painter and church decorator, 1929-34. Assistant to W. Thornton Shiells, medical illustrator, 1934-35. Free-lance medical illustrator, 1937 to the start of World War II. During the war years, was a W.A.A.F. officer responsible for illustrations for publications and records of the R.A.F. Medical Directorate, London.

Media   Watercolor, gouache, ink, pencil.

One-person Exhibition   "Flower Paintings," Foyles Art Gallery, London, 1958.

Collections   Medical Directorate, R.A.F., London; Ashford County Hospital, Middlesex; Royal College of Surgeons, London; British Museum (Natural History), London; Royal Botanic Gardens, Kew; Goulandris Museum, Athens.

Commissions/Works Published in   Ary, S. and Gregory, M. *Oxford book of wild flowers*. London, Oxford University Press, 1960.
Wallis, M., Anderson, E.B., Finnis, V., and others. *Oxford book of garden flowers*. London, Oxford University Press, 1964.*
Brightman, F.H. *Oxford book of flowerless plants*. London, Oxford University Press, 1966.
Wallis, M., Masefield, G.B., and Harrison, S.G. *Oxford book of food plants*. London, Oxford University Press, 1969.
Salt, L. "Plants," *Oxford children's reference library*, London, Oxford University Press, 1972.
Clapham, A.R. *Oxford book of trees*. London, Oxford University Press, 1975.
Plant ecology wall charts (3 out of 5 series, 5 charts each), and *Culinary herbs* (chart), London, British Museum (Natural History), 1972-76.
Illustrations for numerous English conservation publications, journals, calendars, and children's books.

197   "Early Flowers of Chalk Down-Land" Watercolor; 19¾ x 28½"

198   Rose and Firethorn fruits* Watercolor; 8¼ x 6½"

199   South African Daisies and Cape Marigold* Watercolor; 8¾ x 6⅜"

Gifts of the artist

197

NIČOVÁ-URBANOVÁ, Věra

Born   Prague, Czechoslovakia,
11 February 1939.

Address   Letná 1037-54301
Vrchlabí, Czechoslovakia.

Education   Arts and Crafts School,
Prague: studied teaching, 1957.
Pedagogical Institute, Hradec
Králové: B.A., 1964.

Medium   Watercolor.

Career   Aesthetic education
instructor, People's School of Arts, Vrchlabí.

One-person Exhibitions   Vrchlabí, 1971, 1976.

Works Published in   *Živa*, 1964, 1966, 1976.
Příhoda. *Léčivé rostliny (Medical Plants)*. Prague, 1973.
Poster of medicinal plants for the Krkonoše National Park,
1976.

200 Helleborine, *Epipactis palustris*
Watercolor; 10 x 7″

201 Sedge, *Carex atrata*
Watercolor; 10 x 6½″

202 Rock Bramble, *Rubus saxatalis*
Watercolor; 10 x 7½″

203 *Arnica montana*
Watercolor; 9½ x 6″

202

NORMAN, Edward d'Aubigny and Marcia Gaylord

Born   EN: Toronto, Canada,
19 March 1911.
MN: Holyoke, Massachusetts,
21 March 1915.

Address   180 Stage Harbor Road,
Chatham, Massachusetts 02633.

Education   EN: Central Technical
School, Toronto: 1927-31.
MN:Museum School of Fine Arts,
Boston: 1933-36.

Career   Husband-wife illustration team.

Media   Pencil, ink, watercolor.

Collections   MN: Morton Arboretum, Lisle, Illinois.

Group Exhibitions   MN: Hunt Institute International, 1968;
1st International Exhibition of Botanical Illustration;
Johannesburg, 1973; North American Mycological Society,
Dartmouth College, 1975.

Works Published in   Illustrations by Edward and Marcia
Norman:
Hay, J. and Farb, P. *Atlantic shore.* New York, Harper &
Row, 1966.

Ogburn, C., Jr. *Winter beach.*
William Morrow & Co., 1966.

Hay, J. *Sandy shore.* Chatham
Press, Inc., 1968.

Kingsbury, J.M. *Rocky shore.*
Chatham Press, Inc., 1970.

Kingsbury, J.M. *Seaweeds of Cape
Cod and the islands.* Chatham
Press, Inc., 1969.*

Norman, E. and M. *Twelve Cape
Cod drawings.* Chatham Press,
Inc., 1971.

Pond, B. *A sampler of wayside herbs.* Chatham Press, Inc.,
1974.

Gates, D.A. *Seasons of the salt marsh.* Chatham Press, Inc., 1975.
Illustrations by Marcia Norman:

Hubbell, H.W. *Treasures of the shore: a beachcomber's
botany.* Chatham, Massachusetts, Chatham Conservation
Foundation, 1963.

Petry, L.C. and Norman, M.G. *A beachcomber's botany.*
Chatham, Massachusetts, Chatham Conservation
Foundation, 1968.

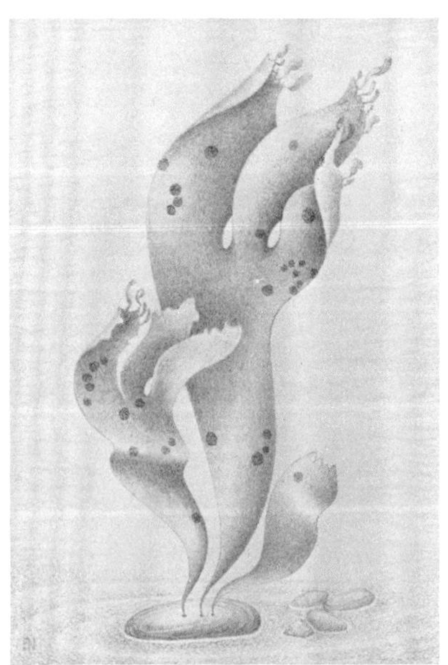

Seaweed illustrations by
Edward Norman:

204 *Ascophyllum nodosum* *

205 *Laminaria digitata* *

206 *Myrionema strangulans* on
Rhodymenia palmata *

Seaweed illustrations by
Marcia Norman:

207 *Isthmoplea sphaerophora* on
Fucus sp. *

208 *Leathesia difformis*

209 *Sphacelaria cirrosa* on *Fucus* sp.*

All pencil; 7⅛ x 5″
Gifts of the artists

206

208

OHTA, Yoai

Born   Aichi Pref. Tahara-Cho, Japan, 30 September 1910.

Address   Naka 1-11-5, Kunitachi-shi, Tokyo 186, Japan.

Education   Studied oil painting under Bunjiro Hosoi, 1926. Manchurian School of Education, Department of Botany: studied botanical illustration under the guidance of Professor Dr. Ichiro Ohga, 1929-30.

Career   Botanical artist and illustrator.

Media   Watercolor, ink, oil, suibokuga (brush and ink monochrome).

Group Exhibitions   Shinjuku Station Building Gallery, 1965; Botanical Art Exhibition, Shinjuku Odakyu Halc Gallery, 1971-76.

Works Published in   Utako Ohga. *Manshu sumire atlas.* Ikimono Shumino Kai, Vol. I, 1931.
Satoru Kurata. *Illustrations of important forest trees of Japan.* Tokyo, Chikyu Sha Co., Ltd., Vol. 2,3,5, 1968.
Eiichi Asayama. *Ornamental plants in colour.* Heibon Sha Ltd., 1971.
Fumio Maekawa. *The wild orchids of Japan in colour.* Tokyo, Seibundo Shinkosha Ltd., 1971.
Jisaburo Ohwi. *Flowering cherries of Japan.* Tokyo, Heibonsha Ltd., 1973.
Ohta, Yoai. *Shokubutsu ga no egaki kata.* Tokyo, Atelier Shuppansha, 1974.

210 Japanese Flowering Cherry, *Prunus serrulata* 'Lannesiana fasciculata'
Watercolor; 13½ x 9½"

211 Sargent or North Japanese Hill Cherry, *Prunus sargentii*
Watercolor; 23½ x 18¼"

Both lent by the artist

211

PAHL, Marion

Born   Chicago, Illinois, 4 April
1926.

Address   5 Fall Street,
Williamsport, Indiana 47993.

Education   Art Institute of
Chicago: B.F.A., 1948; B.A., Art
Education, 1950.

Career   Free-lance botanical artist
and scientific illustrator. Presently
developing a botanical garden of
flora of Warren County, Indiana.

Media   Ink, pencil, oil.

Collections   Field Museum, Chicago; Field Enterprises;
Encyclopaedia Britannica; private collections.

Commissions/Works Published in   *The Encyclopaedia
Britannica.*
The World book encyclopedia.
World history. Laidlaw Brothers.
Field Museum publications.*
Numerous murals and exhibits for the Field Museum.
Exhibits for the Chicago Academy of Science.
Paintings for the Brookfield Zoo.

212 *Uncaria tomentosa* *
Ink; 10¼ x 13¾"

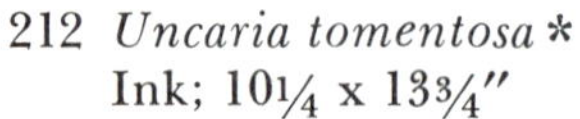

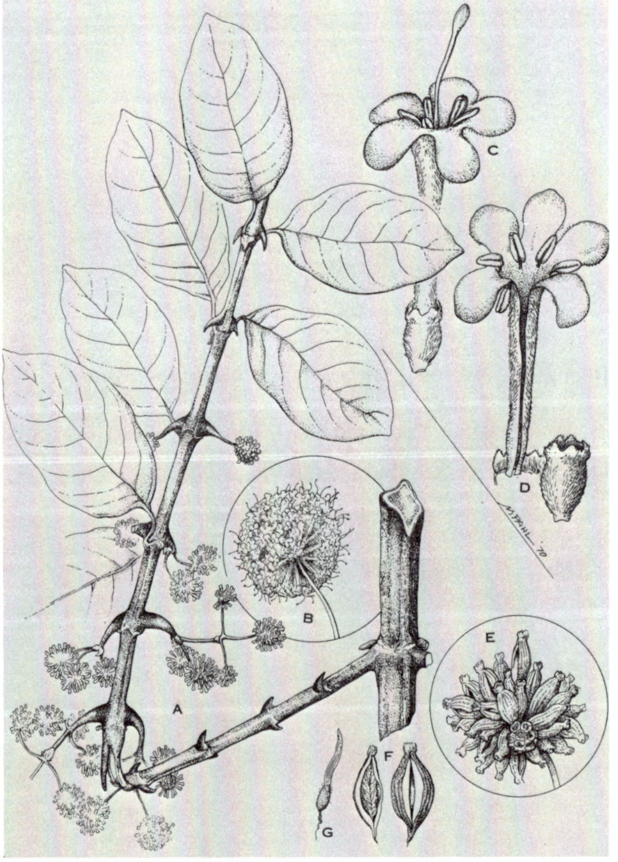

PETERSON, Norton

Born   Harlan, Iowa, 28 April 1916.

Died   Pittsburgh, Pennsylvania,
17 May 1975.

Education   Art Institute of
Pittsburgh, 1937-39.

Career   Painter, printmaker, and
sculptor, 1945-75. Artist and
illustrator, specializing in
industrial subjects, Town Studios,
Pittsburgh, 1955-75. Member of
American Watercolor Society.
Proprietor of the Fragment Press (cited in Roderick Cave's
Private press books of 1970. Middlesex, 1972).

Media   Ink, pencil, watercolor, etching, woodcut.

One-person Exhibitions   Frick Fine Arts Department,
University of Pittsburgh; University of Pittsburgh Book
Center, 1966; Old Post Office Museum, Pittsburgh History
and Landmarks Foundation, 1972.

Group Exhibitions   American Watercolor Society, New York;
Pittsburgh Playhouse print show, Pittsburgh; Butler Art
Institute, Youngstown, Ohio, 1966; Hunt Institute
International, 1972, and Homegrown Exhibition, 1975;
Hunt Institute Travel Shows (International); Watercolor
U.S.A., Springfield Art Museum, Springfield, Missouri, 1975.

Awards   First Prize, Pittsburgh Playhouse print show;
Purchase Prize, Butler Art Institute, Youngstown, 1966.

Collections   Butler Art Institute, Youngstown; Pittsburgh
History and Landmarks Foundation, Pittsburgh; private
collections.

Works Published in   Cover of the Hunt Institute Year-end
Report, 1969.
Advertising illustrations in magazines and trade journals.

213   *Rhododendron* hybrid
Watercolor; 21⅝ x 17⅝″

PINGITORE, Eugenio José

Born   Buenos Aires, Argentina, 8 June 1947.

Address   R.F.I. Burela 3519, Villa Urquiza, Buenos Aires, Argentina.

Education   Escuela Cristobal, Buenos Aires: Technician in Botany and Gardening.

Career   Free-lance botanical artist. Head of Taxonomy, Instituto de Botánica, Buenos Aires. Botanist, Relaciones Fitologicas Intercontinentales, Buenos Aires. Professor of Tropical Botany, Buenos Aires Professional and Cultural Center. Conducts a series of television programs on botany. Collects South American tropical plants for taxonomic study by many international botanical institutions and herbariums. Formerly, founder of *Palmetum* landscape project, Buenos Aires, 1973. Botanist, designer, and consultant, Sociedad Argentina de Horticulture, Buenos Aires, 1974-76. Curator of Cultivated and Indigenous Plants Section, Carlos Thays Botanical Garden, Buenos Aires, 1976.

Media   Ink, watercolor.

One-person Exhibition   Permanent exhibition in the Carlos Thays Botanical Gardens, Buenos Aires.

Group Exhibitions   Tropical Palms of the World, Atlántida Bookstore, Buenos Aires, 1974; First Flower International Exposition, Bogotá, Colombia, 1974.

Honors   Honorable Mention, Sociedad Argentina de Horticultura, 1976; Honorary Member, Sociedad Argentina Científica.

Works Published in   Pingitore, E. *Plant notes from Tecnico Botanico.* American Association of Botanical Gardens and Arboreta, 1974.
Pingitore, E. *The Republic of Argentina tree ferns.* Los Angeles International Fern Society, 1976.
Pingitore, E. "Hunting bromeliads in the Argentine Republic," *The Bromeliad Society bulletin,* Vol. XXVI, 1976.
Dimitri, M. *Enciclopedia Argentina de horticultura y jardineria.* 1977.
Principes, The Palm Society, 1975.
The Bromeliad Society bulletin, 1975-76.
Sociedad Argentina de horticultura, 1972-77.
Revista del Instituto Municipal de Botanica [in progress].

214   Palm, *Chamaedorea* sp.
Ink; 22⅝ x 18⅜"

215   Argentine Rain Forest, Misiones Province
Ink; 23¼ x 18"
Both lent by the artist

214

PISTOIA, Marilena

Born   Milan, Italy, 31 December 1933.

Address   Via Azzurra 16, Bologna, Italy.

Education   Artistic High School, Monza: diploma, 1951. Academy of Fine Arts, Milan: degree, 1955. Anatomical Design School, Bologna: diploma, 1957.

Career   Free-lance botanical illustrator.

Media   Watercolor, ink, oil, etching.

Collections   Ramo Editoriale degli Agricoltori, Rome; Arnoldo Mondadori, Publisher, Verona.

Works Published in   Goidanich, G. (ed.). *Avversità della piante agrarie*. Rome, R.E.D.A., Vols. I-V, 1958-65.
Bianchini, F. "Higher plants," *Monographic encyclopedia of natural sciences*, Verona, Arnoldo Mondadori, 1971.
Corbetta, F. "Lower plants," *Monographic encyclopedia of natural sciences*, Verona, Arnoldo Mondadori, 1971.
Bianchini, F., and Corbetta, F. *The complete book of fruits and vegetables*. New York, Crown Publishers, Inc., 1975 (*I frutti della terra*. Verona, Arnoldo Mondadori, 1973).
Bianchini, F. and Corbetta, F. *Le piante della salute*. Verona, Arnoldo Mondadori, 1975.

216   Onions, *Allium* sp.
Watercolor; 10⅞ x 15″

217   Scabiosa, Pincushion Flower, *Scabiosa* sp.
Watercolor; 7¾ x 7⅞″
Lent by the artist

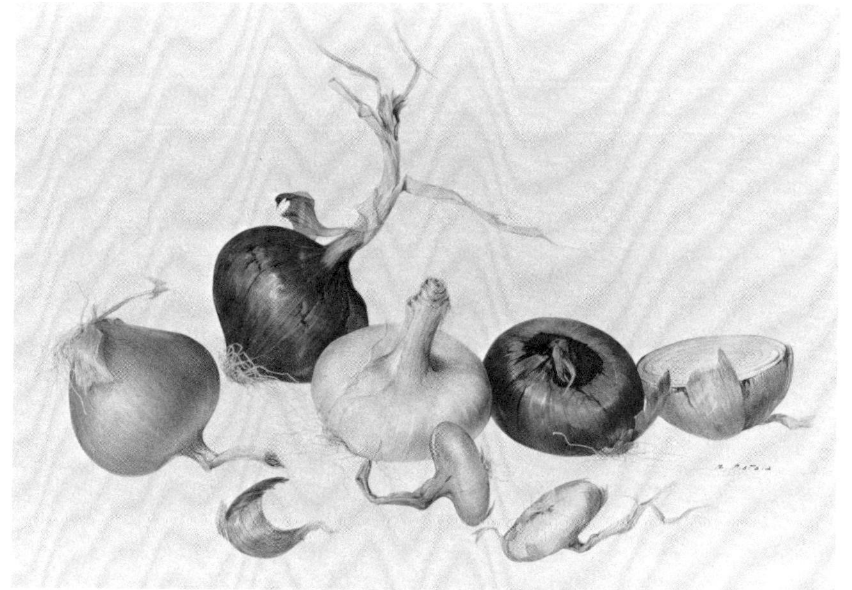

216

POMEROY, Mary Barnas (Mrs. Fred G.)

Born   Frankfurt am Main, Germany, 3 March 1921.

Address   88 Boronda Road, Carmel Valley, California 93924.

Education   Colegio 24 de Mayo, Quito, Ecuador: 1936-38. Studied with landscape painter Carl Barnas (father), Quito: 1938-44. Pennsylvania Academy of Fine Arts, Philadelphia: studied painting with Daniel Garber, Roy Nuse, and Franklin Watkins, 1946-48.

Career   Free-lance artist, illustrator and writer, 1953 to present. Formerly, illustrator for: Geology Department, Universidad Central, Quito, 1939-42; Paleontology Department, International Petroleum Co., Guayaquil, Ecuador, 1943-44; Botany Department, University of California, Berkeley, 1948-53.

Media   Ink, watercolor, oil.

One-person Exhibitions   Universidad Central, Quito, 1940; Academy of Natural Sciences, Philadelphia, 1946; Academy of Sciences, San Francisco, 1948; University of California, Berkeley, 1949; Carmel Valley Center, California, 1966; County Library, Monterey, California, 1969; Public Library, Houston, 1969; Museum of Natural History, Pacific Grove, California, 1970; Carmel, California, 1972; Missouri Botanical Garden, St. Louis, 1974-75; Indiana University, Bloomington, 1975; Clark Museum, Scripps College, Claremont, California, 1976.

Group Exhibitions   Cleveland Museum of Art, 1966; Hunt Institute Internationals, 1964, 1972.

Collection   University of California, Berkeley.

Works Published in *Espinosa* (a botanical textbook). Quito, 1938.
Davidson, J.F. *The genus Polemonium.* Berkeley, University of California Press, 1950.

Egerod, L.E. *An analysis of the siphonous Chlorophycophyta.* Berkeley, University of California Press, 1952.
Mason, H.L. *A flora of the marshes of California.* Berkeley, University of California Press, 1957.
Metcalf, W. *Native trees of the San Francisco Bay area.* Berkeley, University of California Press, 1959.
Pomeroy, M.B. "Saloya," *Américas,* 1967.
Pomeroy, M.B. " Ecuador's edible jewels," *Américas,* 1969.
Pomeroy, M.B. "Pasochoa and Cotopaxi," *Américas,* 1971.
Carmel Valley outlook, October, 1972.
Poetry shell, 1972-75.

219

218 Beggar-tick, *Bidens pilosa* var. *radiata*
Watercolor; 8¼ x 5¼″

219 *Steiractinia sodiroi*
Watercolor; 10 x 6½″

220 *Palicourea* sp.
Watercolor; 9¾ x 5¾″

POOLE, Monica

Born   Canterbury, England,
20 May 1921.

Address   67 Hadlow Road,
Tonbridge, Kent, England.

Education   Central School of Art
and Design, London: Diploma,
1949.

Career   Free-lance artist and
designer of book jackets and
greeting cards.

Medium   Wood engraving.

One-person Exhibition   Trinity Gallery, Colchester, 1970.

Group Exhibitions   Royal Academy; Royal Glasgow Institute
of Fine Arts; Society of Wood Engravers; Royal Society of
Painters, Etchers, and Engravers; 2nd International Biennial
of Prints, Pescia, Italy, 1968; Hunt Institute International,
1972; Hunt Institute Travel Shows (Prints).

Honors   Fellow, Royal Society of Painters, Etchers, and
Engravers. Member, Art Workers Guild.

Collections   Pistoia Art Gallery and Museum; Essex, Kent, and
Lancashire Education Committees; South London Art
Gallery; Victoria and Albert Museum, London; Ashmolean
Museum, Oxford; Hunterian Museum, Glasgow; Museum
Boymans van Beunengen, Rotterdam; Newport Museum and
Art Gallery; Portland State University; Boston Public Library.

Commissions/Works Published in   Turnor, R. *Kent*. London,
Paul Elek, 1950.
Hadfield, J.C. (ed.) *Saturday book no. 31*. London,
Hutchinson's Publishing Co., 1971.
Book Jackets and Cover Designs for:
Macself, A.J. *Ferns for garden and greenhouse*. London,
W.H. & L. Collingridge, 1950.
Christmas booklist for Methuen & Co., London, 1950.
Morton, John B. *Camille Desmoulins and other stories of
the French Revolution*. London, Werner Laurie, 1950.
Book of wildflowers. Nos. 1-4, London, W.R. Chambers, Ltd.,
1953.
Amateur gardening diary. London, W.H. & L. Collingridge,
1953-55.
Beckford, Peter. *Thoughts on hunting*. London, Methuen,
1965.

221 "Water Lilies," *Nymphaea* sp.
Wood engraving; 4⅞ x 7¼"
Gift of the artist

222 *Magnolia* sp.
Wood engraving; 7¾ x 6¾"

221

POVALL, Lois Martin

Born Somerset, England, 17 February 1905.

Address 11 Holland Place, Essenwood Road, Durban, 4001 Natal, Republic of South Africa.

Education Penrhôs College, Colwyn Bay, North Wales: 1917-23. Liverpool College of Art: 1924-25. Regent St. Polytechnic Art School, London: textile design, 1927-29. Studied art in Florence. Self-taught botanical artist.

Career Free-lance botanical artist. Formerly, instructor of Royal Drawing Society's syllabus in several private schools, Devon, England, 1938-43. Settled in South Africa, 1951.

Medium Watercolor.

One-person Exhibitions Neil Sack Gallery, Pietermaritzburg; Walsh Marais Gallery, Durban (three showings).

Group Exhibitions Botanic Artists of South Africa, Pietermaritzburg, 1967; International Exhibition of Botanical Art, Johannesburg, 1973; South African Association of Arts, Pretoria, 1974.

Collections The Campbell Africana Collection, University of Natal, Durban; private collections.

Commissions Medical drawings for surgeons; series of illustrations depicting the life cycles of internal parasites in man, University of Natal, Durban, 1966-67; two botanical paintings for the Campbell Africana Collection, University of Natal, 1975; numerous botanical studies for private collectors, since 1952.

Works Published in Rosenthal, Eric. *Encyclopaedia of southern Africa*, London, Warne, 1961. "*Haemanthus magnificus*," *Veld and flora*, 1972.

223 Coral Tree, *Erythrina latissima* (flower)
Watercolor; 10½ x 14″

224 Coral Tree, *Erythrina latissima* (fruit)
Watercolor; 9½ x 4¾″

225 *Crinum macowanii*
Watercolor; 15⅜ x 11″

Gifts of the artist

223

POWELL, Linda Kay

Born Los Angeles, California, 4 July 1943.

Address 6 Sutherland Place, Manitou Springs, Colorado.

Education Glendale Junior College, Glendale, California: A.A. Art Center College of Design, Los Angeles: B.F.A.

Career Free-lance illustrator and designer. Formerly, Assistant Art Director, 1972-75, and Art Director, 1975-76, Looart Press Inc., Colorado Springs.

Media Watercolor, gouache, ink, acrylic, charcoal.

One-person Exhibitions Whites Gallery, Montrose, California, 1968, 1969, 1970, 1972; George Nix Gallery, Colorado Springs, 1975, 1977.

Group Exhibitions George Nix Gallery, 1975, 1976, 1977; Gallery 323, Casper, Wyoming, 1976, 1977; Turkey Creek Gallery, Morrison, Colorado, 1975, 1976, 1977; Simpatico Gallery, Houston, Texas, 1976, 1977.

Commissions/Works Published in Calendars, books, and greeting cards produced by Looart Press Inc., 1968-77. *

Steiner, B. *Biography of a desert bighorn.* New York, G.P. Putnam's Sons, 1975.

Evarts, S. *The art and craft of greeting cards.* Westport, Connecticut, Northlight Publications, 1975.

Schaefer, J. *An American bestiary.* Boston, Houghton Mifflin Co., 1975.

Coerr, E. *Biography of a kangaroo.* New York, G.P. Putnam's Sons, 1976.

Steiner, B. *Biography of a kangaroo rat.* New York, G.P. Putnam's Sons, 1977.

The rabbit is next. Racine, Wisconsin, Western Publishing Co., 1977.

Scat, scat. New York, Platt & Munk, 1977.

Social expression designs for George Good Corporation, Covina, California.

Illustrations for *Ranger Rick* (magazine), National Wildlife Federation, Washington, D.C.

Needlework designs for Dimensions Inc., Reading. Pennsylvania, and for Columbia-Minerva Corporation, New York.

226 Sycamore, *Platanus* sp. *
Watercolor; 4 x 5″

227 Big Leaf Maple, *Acer macrophyllum* *
Watercolor; 4 x 5″

228 Northern Red Oak, *Quercus rubra* *
Watercolor; 4 x 5″

229 Sweet Gum, *Liquidambar styraciflua*
Watercolor; 4 x 5″

230 "Bee and Berries" *
Watercolor; 5¼ x 7¼″

Gifts of Looart Press, Inc.,
Colorado Springs, Colorado

230

PRENTISS, Thomas Saunders

Born   Buffalo, New York,
13 September 1926.

Address   66 Golden Horn Road,
Oakdale, New York 11769.

Education   Albright Art School,
Buffalo: 1946-47. Art Students
League, New York: 1949-50.
Apprentice to Corrado Cagli,
1947-49. Self-trained scientific
illustrator.

Career   Scientific illustrator,
Scientific American, 1960 to present. Formerly, easel
painter, until 1960.

Media   Oil, egg tempera, pencil, ink, silverpoint, goldpoint.

One-person Exhibitions   Durlacher Brothers, New York, 1954,
1957.

Group Exhibitions   Western New York Annuals, 1952-54;
Rhode Island School of Design, 1955.

Awards   Prizes for oils and egg temperas, Western New York
Annuals, 1952-54.

Collection   Whitney Museum, New York.

Works Published in   *Art News*, 1954.
Many illustrations for *Scientific American*, 1960 to present.*

231 Hemp, Marijuana, *Cannabis sativa**
Ink; 16⅜ x 12¼"

PRINCE, Martha (Mrs. Jordan H.)

Born   Birmingham, Alabama, 8 August 1925.

Address   9 Winding Way, Locust Valley, New York 11560.

Education   Studied art from childhood with mother, Martha Fort Anderson, former Head of the Art Department, University of Alabama. Piedmont College, Georgia: B.A., 1944. Art Students League, New York: studied under Reginald Marsh and Howard Trafton, 1944-45.

Career   Free-lance botanical artist, writer, and lecturer specializing in rhododendrons and azaleas. Formerly, artist and illustrator for the United States Army, 1945. Art director, Mears Advertising, New York, 1945-48.

Media   Watercolor, ink, pencil, tempera.

One-person Exhibitions   Horticultural Society of New York, 1972; Callaway Gardens, Georgia, 1972; Planting Fields Arboretum, Long Island, New York, 1973; Bayard Cutting Arboretum, Long Island, 1974; Seaford Public Library, 1974; Islip Town Gallery, Long Island, 1974, 1977; United States National Arboretum, 1978.

Group Exhibitions   Numerous, including: Maryland Art Institute, Baltimore; Delgado Musem, New Orleans; County Art Gallery, Locust Valley, New York, 1974.

Collections   Seaford Library, Long Island: Hofstra University Library, Long Island; Planting Fields Arboretum, Long Island; many private collections.

Works Published in   Prince M. "Rhododendrons and azaleas," *American Rhododendron Society quarterly,* 1969.
Prince, M. "Rhododendrons: propagating and hybridizing," *American Rhododendron Society quarterly,* 1969.
Prince, M. "Color range in *Rhododendron calendulaceum,*" *American Rhododendron Society quarterly,* 1973.
Prince, M. "The elusive *Shortia,*" *American Rock Garden Society bulletin,* 1973.
Prince, M. "Sky-paint, flower of the Cherokee," *Atlanta journal and constitution Sunday magazine,* 1973.
Wildflowers of America. Nashua, New Hampshire, Doehla, Inc., 1974.
Wildflower notes. Long Island, Aboretum Press, 1975.
Flowers of the wild. Denmark, George Caspari, 1976.
Cover illustrations for Planting Fields Arboretum *Newsletter,* and *American Rhododendron Society quarterly.*
Illustrations for *New York times; Harper's bazaar; Life;* Cutting Arboretum, Long Island, publications; and many other magazines. Illustrations and articles for *Horticulture, American horticulturist,* and *Living wilderness.*

232   Oconee-bells, *Shortia galacifolia*
Watercolor; 5 x 7"

233   Bloodroot, *Sanguinaria canadensis*
Watercolor; 5 x 7"

234   Franklin Tree, *Franklinia alatamaha*
Watercolor; 5 x 7"

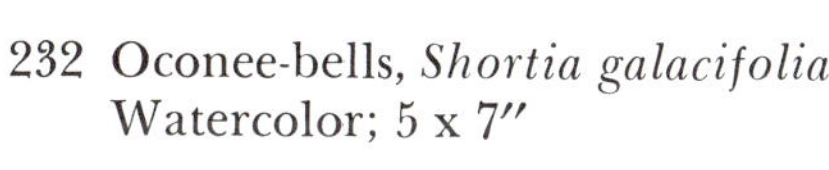

234

RAVINDRAN, O. T.

Born   Cannanore (Kerala State), India, 27 June 1934.

Address   59 Nandanam Extension, Madras 600 035, India.

Education   Government Arts College, Madras: B.A., Botany, 1956. Madras Christian College, Madras: M.A., Philosophy, 1958. Self-taught botanical artist.

Career   Free-lance journalist, photographer, botanical artist, and landscape designer. Founder (1976) and proprietor of Garuda Gardens, Madras. Formerly, press liaison officer, British Information Services, Madras, 1958-67.

Medium   Watercolor.

One-person Exhibitions   Orchid paintings, Seminary Hall, Local Library Authority Buildings, Madras, 1967; Orchids and Trees, Sarala Art Center, Madras, 1970; Plants I Love, Sarala Art Center, Madras, 1974; Orchids, Massachusetts Horticultural Society, Boston, 1974.

Group Exhibitions   Trees of India, Royal Horticultural Society, London, 1972; Orchids of India, Royal Horticultural Society, London, 1973; Botanical Society of South Africa, 1973; Hunt Institute International, 1972; Climbers in a Madras Garden, Royal Horticultural Society, London and Hotel Taj Coromandel, Madras, 1976; Hunt Institute Travel Shows (International).

Awards   Silver Grenfell Medals, Royal Horticultural Society, 1972, 1973, 1976.

Works Published in   *The Hindu, The Indian Express, The Mail.*

235 Clock-vine, *Thunbergia grandiflora*
Ink and watercolor; 13½ x 10¼″

236 Coral Tree, *Erythrina indica*
Watercolor; 13½ x 13″

Both indefinite loans of the artist

236

RIEMER-GERHARDT, Elisabeth

Born Buitenzorg (Bogor), Dutch East Indies (Indonesia), 14 April 1931.

Address Bloemenlaan 2, Heelsum, Netherlands.

Education Royal Academy of Art, The Hague: Master's degree, Art Teacher's degree.

Career Art Director, N.C.A., 1974-77; art teacher, States Highschool, Tiel, 1969-77. Formerly, botanical illustrator, State Agricultural University, Wageningen, 1954-57. College assistant, State Agricultural University, Wageningen, 1957-59. Actress, 1959-69.

Media Ink, oil, watercolor, pencil.

One-person Exhibitions Laboratorium voor Plantensystematiek, Wageningen, 1966; Leeuwenborgh, Wageningen, 1975.

Group Exhibitions Exposition Internationale des Academies des Beaux Arts, Milan, 1953; The Hague, 1965; Maria Miriam Exhibition.

Honor Member, Dutch Art Directors of Dramatic Art.

Collections State Agricultural University, Wageningen; many private collections.

Works Published in *Het aquarium,* 27th year, No. 7, 1956. *Mededelingen van de Landbouwhogeschool te Wageningen,* 1957. *Mededelingen van het laboratorium vor virologie,* Wageningen, 1957. Numerous publications of the Laboratory of Plant Taxonomy and Plant Geography, Wageningen.

237 Witch Hazel, *Hamamelis japonica*
Watercolor; 11½ x 8″

238 Sargent or North Japanese Hill Cherry,
Prunus sargentii
Watercolor; 12½ x 7¾″

Permanent loans of the Laboratorium voor Plantensystematiek en -Geografie, Landbouwhogeschool, Wageningen, Netherlands

237

RISSLER, Gerd

Born   Stockholm, Sweden,
29 January 1909.

Address   Stopvagen 23,
161 46 Bromma, Sweden.

Education   Konstskolan, Stockholm:
studied sculpture, 1929. School of
Fine Arts: 1930-31. Otte Sköld's
School of Painting: 1932.
University of Stockholm: 1932-33.

Career   Free-lance illustrator.

Media   Ink, watercolor, oil.

One-person Exhibitions   Stockholm, 1955, 1975; Norrtälje,
1966; Strängnäs, 1969.

Group Exhibitions   Stockholm, 1960, 1969, 1970, 1973, 1976;

Hunt Institute International, 1972.

Works Published in   Falck, K., Hammarsten, O., and Friberg, O.
Enhetsskolans biologi. Stockholm, Svenska Bokforlaget-
Bonniers, 1959.
Söderling-Brydolf, C. *Blommorna vid Blå Nilen*. Stockholm,
Natur och Kultur, 1965.
Söderling-Brydolf, C. *Bland Liljor och Lejon*. Stockholm,
Natur och Kultur, 1966.
Linnman, N. and Wennerberg, B. *Grundskolans
naturkunskap*. Stockholm, Magnus Bergvalls Förlag, [1968].
Mejrud, B. and Rune, O. *Växter och Djur*. Stockholm, A.V.
Carlsons, 1969.
Rissler, G. *Svenska växtsamhällen (Swedish plant colonies)*.
13 color posters published by Norstedts Skolboksförlag,
Stockholm, n.d.

239 Betony, *Stachys silvaticus;*
Cynanchum vincetoxicum; Bladder Campion,
Silene vulgaris
Watercolor; 9¾ x 5¾″
Lent by the artist

240 Fumitory, *Fumaria officinalis;*
Large-flowered Hemp-Nettle, *Galeopsis speciosa*
Watercolor; 9¼ x 6″

239

ROESENER, Richard

Born   Fort Wayne, Indiana, 11 November 1946.

Address   889 Clayton Street, San Francisco, California 94117.

Education   Herron School of Art of Indiana University: B.F.A., 1968.

Career   Free-lance illustrator, January 1977 to present. Formerly, substitute art instructor, Indianapolis School District and Metropolitan School District of Wayne Township, Indianapolis, 1968. Industrial arts instructor, Metropolitan School District of Perry Township, Indianapolis, 1969. Scientific illustrator, 1969-72; Chief Scientific Illustrator, 1973-76, Museum of Natural History, Chicago, Illinois. Member of Guild of Natural Science Illustrators, Washington, D.C.

Media   Ink, pencil, watercolor.

One-person Exhibitions   Albert Eisenlau Gallery, New York, 1974, 1975.

Group Exhibitions   Hot Flash of America, San Francisco, 1976, 1977; Jacques Baruch Gallery, Chicago, 1976; Chicago Horticultural Society, 1976.

Award   Katherine Ayeres Smitheron Award for excellence in drawing, 1966.

Collections   Field Museum of Natural History, Chicago; First State Bank of Porter, Porter, Indiana; Playboy Enterprises, Inc., Chicago.

Works Published in   Numerous journals, including: Field Museum of Natural History *Bulletin*, Vol. 41, no. 1, 1970; Vol. 42, no. 2, 1971; Vol. 44, nos. 2,7,8, 1973; Vol. 45, no. 1, 1974.
Fieldiana, Geology, Vol. 20, no. 5, 1970; Vol. 23, nos. 1-3, 1971; Vol. 28, 1972; Zoology, Vol. 58, no. 9, 1971; Botany, Vol. 36, no. 1, 1973; Vol. 37, 1974.
Phycologia, Vol. 10, nos. 2,3, 1971; Vol. 11, no. 1, 1972.
de la Torre, L. *Extrait de mammalia*, Vol. 35, no. 2, 1971.
Traylor, M. *Journal für Ornithologie*, Vol. 112, 1971.
Friedman, A., Olsen, E., and Bird, J. *American antiquity*, Vol. 37, no. 2, 1972.
Scholastic news trails, Vol. 27, no. 4, 1973.
Nitecki, M. *University of Chicago magazine*, Vol. LXV, no. 6, 1973.
Simpson, D. *Phytologia*, Vol. 29, no. 4, 1974.*
DeBlase, A. and Martin, R. *A manual of mammalogy*. Dubuque, William C. Brown Company, 1974.
Dybas, H. and Lloyd, M. *Ecological monographs*, Vol. 44, no. 3, 1974.
Schram, F. and Nitecki, M. *Journal of paleontology*, Vol. 49, no. 3, 1975.
Where it's at—New York City, 1975.
Oui, Vol. 5, no. 12.

241 Royal Fern, *Osmunda regalis* var. *spectabilis;* Cinnamon Fern, *Osmunda cinnamomea*
Ink; 15⅝₁₆ x 11¾″

242 *Paullinia* sp. *
Ink; 13¼ x 8¹¹₁₆″

243 *Paullinia* spp.*
Ink; 7½ x 11⅜″

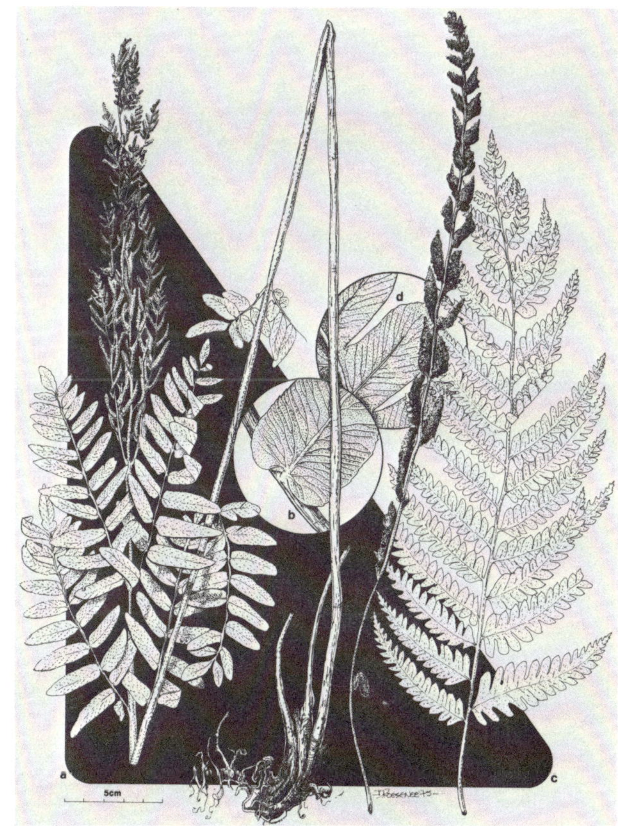

241

SAITO, Charles Manabu

Born   Tokyo, Japan, 12 May 1929.

Address   Box 200, Stillwater, New Jersey 07875.

Education   Pratt Institute, New York: B.A., Industrial Design, 1957. Self-taught botanical artist.

Career   Free-lance botanical artist.

Medium.   Watercolor.

One-person Exhibitions   Brooklyn Botanic Garden, 1967, 1968; New York Botanical Garden, 1968; New York Horticultural Society, 1970; Frame House Gallery, Louisville, Kentucky, 1970.

Group Exhibition   Hunt Institute International, 1972.

Commissions/Works Published in   "Cactus," *Audubon magazine*, 1974.
Vinning, F.D. *The golden nature guide on cacti*. New York, Western Publishing Company, Inc., 1974.
"A gallery of nature art," *Audubon magazine*, 1974.
Day, J.W. *What is a tree?* New York, Western Publishing Company, Inc., 1975.
Webster, V. *From one seed*. New York, David McKay Publisher, 1976.
Flower paintings for National Wildlife Federation wildlife stamp series.
Collector print series, Frame House Gallery, Louisville, Kentucky. *
Christmas card designs for *Audubon magazine* and American Artist Group Inc.

244 "Maidenhair Fern," *Adiantum pedatum**
Watercolor; 19¾ x 23¾"
Lent by the artist

SANTOS, Norga Maria Mascarenhas dos

Born   São Paulo, Brazil, 28 July 1946.

Address   Instituto de Botânica, Caixa Postal 4005/01000, São Paulo - SP, Brazil.

Education   Fundação Armando Alvares Penteado: Plastic Arts, 1975. Brazil-Japan Cultural Union: watercolor, 1977.

Career   Botanical Illustrator, Institute of Botany, São Paulo, Brazil.

Medium   Ink.

Group Exhibition   Hunterdon Art Center, Clinton, New Jersey, 1977.

Collection   Instituto de Botânica.

Works Published in   *Rickia* and *Hoehnea*.

245 Glory Bush, *Tibouchina holosericea*
Pencil; 17½ x 9½"
Lent by the artist

SCHAFER, Alice Pauline

Born Albany, New York, 11 February 1899.

Address 33 Hawthorne Avenue, Albany, New York 12203.

Education Albany School of Fine Arts, Albany, New York: course of study in art, graduated 1919. Self-taught printmaker.

Career Free-lance printmaker. Formerly, Staff Illustrator and Technician, Albany Medical College, New York, 1930's - 1940. Began printmaking career, 1940. Membership in: Print Club of Albany, New York (president); National Arts Club; Southern Vermont Artists; Society of American Graphic Artists; North American Mycological Association; Pen and Brush.

Media Linocut, wood engraving, etching.

One-person Exhibition Albany Institute of History and Art, New York, 1962.

Group Exhibitions Numerous, including: Albany Institute of History and Art; Library of Congress, Washington, D.C.; National Academy of Design, New York; Pennsylvania Academy of Fine Arts, Philadelphia; Butler Institute of American Art, Youngstown, Ohio; North American Mycological Association, Dartmouth College, 1975; Hunt Institute International, 1972; Hunt Institute Travel Shows (Prints).

Awards Numerous, including: Gold Medal, 1966, and First Prize, Etching, American Artists Professional League; Alice Standish Buell Memorial Prize, Pen and Brush, New York; John Taylor Arms Memorial Purchase Prize, Print Club of Albany; Second Prize, Rensselaer County Council of the Arts.

Collections Numerous, including: Metropolitan Museum of Art, New York; Pennsylvania Academy of the Fine Arts, Philadelphia; Boston Public Library; New York Public Library; Print Club of Albany; Southern Vermont Art Center; Butler Institute of American Art.

Commissions/Works Published in *Cancer research,* 1947. *American journal of anatomy,* Vol. 85 (2), 1949. Prints for the National Commercial Bank and Trust Company, and the Print Club of Albany.

246 Mulberries, *Morus* sp.
Color linocut; 6¾ x 11"

247 Gooseberries, *Ribes* sp.
Color linocut; 6¾ x 11"

248 "Indeed, No!" Mushroom,
Amanita pantherina
Color linocut; 9½ x 9"

246

SCHEEPMAKER, W.

Born   Utrecht, Netherlands,
7 May 1911.

Address   Doorwerth, Dillenburg
350, Netherlands.

Education   Kunstacademie
Artibus, Utrecht.

Career   Botanical artist,
Department of Botany, State
Agricultural University,
Wageningen, and the
International Association for
Plant Taxonomy and Nomenclature, Utrecht. Formerly,
illustrator, Botanical Museum and Herbarium, Utrecht.

Medium   Ink.

Works Published in   Students' books in Utrecht, Amsterdam-
Groningen, Leiden, and Wageningen.

249 *Strychnos johnsonii; S. samba*
Ink; 13¼ x 8⅞″

250 Apple, *Malus prattii*
Ink; 11 x 8″

Permanent loans of the Laboratorium voor
Plantensystematiek en -Geografie,
Landbouwhogeschool, Wageningen, Netherlands

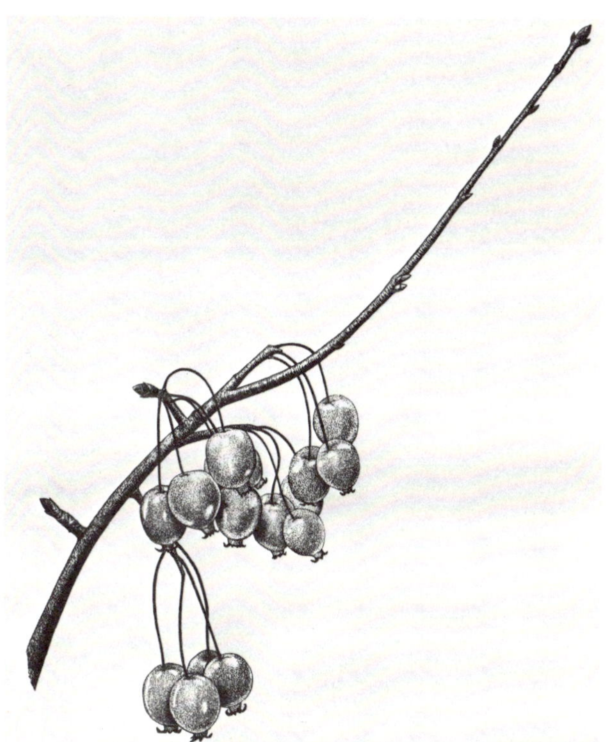

250

SCHROEDER, Jack Rankin

Born Poughkeepsie, New York, 13 September 1927.

Address Box 98, Still Pond, Maryland 21667.

Education Self-taught artist.

Career Free-lance botanical and zoological artist. President, Schroeder Prints, Inc.

Media Ink, pencil, watercolor.

Group Exhibition Hunt Institute International, 1972.

Award 4th Annual Maryland Migratory Waterfowl Contest.

Works Published in Hale, M.E. *The lichens.* Dubuque, Iowa, William C. Brown Co., 1969.

Clyde, F., Roper, E. and Mangold, M. "A new species of *Illex*," *Proceedings of the Biological Society of Washington,* Vol. 82, 1969.

Sawyer, J.O. and Lindsay, A.A. *Vegetation of the life zones in Costa Rica.* Indianapolis, Indiana Academy of Science, 1971.

Holdridge, L.R., Grenke, W., Hathaway, W.B., Tosi, J., and Liang, T. *Forest environment in tropical life zones.* Oxford, Pergamon Press, 1971.

Kornicker, L.S. "Myodocopid Ostracode," *Smithsonian contributions to zoology,* No. 39, 1970.

King and Barroso. "New taxa of Compositae (Eupatorieae) from Brazil," *Brittonia,* Vol. 23, 1971.

Humphrey, Bridge, Reynolds, and Peterson. *Birds of Isla Grande (Tierra Del Fuego).* Washington, D.C., Smithsonian Institution, 1971.

Stephens, J.S., Jr. and Springer, V.G. *"Neoclinus nudus,"* *Proceedings of the Biological Society of Washington,* Vol. 84, 1971.

Springer, V.G. "Revision of the fish genus *Ecsenius*," *Smithsonian contributions to zoology,* No. 72, 1971.

King, R.M. and Robinson, H. "The genera of the Eupatorieae (Asteraceae)," *Smithsonian contributions to botany.*

Wylie, S.R. and Furlong, S.S. *Key to North American waterfowl.* Philadelphia, Livingston Publishing Company, 1972.

Norstog, K. and Long, R.W. *Plant biology.* Philadelphia. W.B. Saunders Company, 1977.

251 *Neomirandea guevarii*
 Opaque watercolor; 16⅝ x 13½"

SCHROER, Elisabeth

Born   Zürich, Switzerland, 26 May 1948.

Address   Atelier für Wissenschaftliches Zeichnen, Rosenbergstrasse 13, CH-8555, Müllheim, Switzerland.

Education   Gymnasium, Zürich: 1962-68. Kunstgeweibeschule, Zürich: 1968-71. Studied scientific illustration under Karl Schmid, Gockhausen, Switzerland.

Career   Teacher at the Kunstgeweibeschule, St. Gall, Switzerland.

Media   Watercolor, pencil, etching.

One-person Exhibitions   Galerie Burkharbhof, Egnach, Switzerland, 1970, 1973; "Klein Galerie," Bad-Waldsee, West Germany, 1975; Neufeld-Galerie, Lustenau, Austria, 1975; Laudenberg-Gesellschaft Schloss, Arbon, Switzerland, 1977.

Group Exhibitions   Zürcher Künstler im Helmhaus, Zürich, 1970, 1971, 1972; Thurgauer Künstlergruppe: Buchs SG, 1974; Müllheim, 1974, 1975; Frauenfeld, 1976.

Collections   Ciba-Geigy AG, Basel; Hotel Baur au Lac, Zürich; Firma E. Bruderer Maschinenfabrik (Arbon, Switzerland), Huntsville, Alabama.

Works Published in   *Ciba-Geigy-Unkrauttafeln*. Basel, Ciba-Geigy AG, 1968-70, 1976-78.
Häfliger, E. *Schweizer Illustrierte*. Zofingen, Switzerland, Ringier-Verlag, 1970.
Tages-Anzeiger. Zürich, 1973.
Dosch, P. *Introducción à la terapia neural con anestésicos locales*. Fieberbrunn, Austria, Gebro G. Broschek AG, 1974.

252 Carrot, *Daucus carota* var. *sativus*
Watercolor; 12 x 9″

253 "Pinot Chardonnay," Wine grape,
Vitis vinifera cv.
Watercolor; 16 x 11⅜″
Lent by the artist

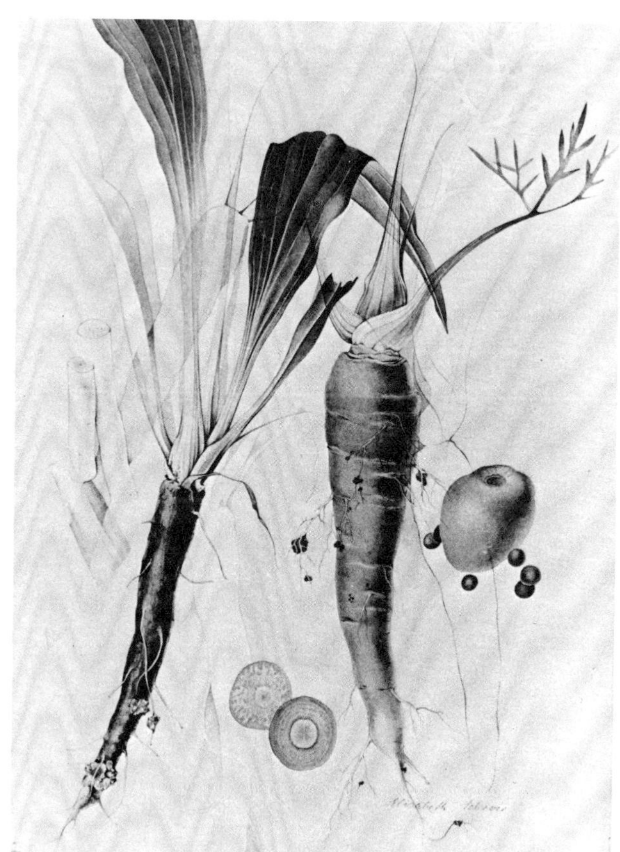

252

SCHWARTZ, Carl E.

Born   Detroit, Michigan, 20 September 1935.

Address   4228 North Hazel Street, Chicago, Illinois 60613.

Education   Art Institute of Chicago: B.F.A., 1957. University of Chicago: B.F.A., 1957.

Career   Painter. Instructor of figure painting and drawing, Suburban Fine Arts Center, 1960 to present, and North Shore Art League, 1958 to present.

Media   Acrylic, lithography.

One-person Exhibitions   Numerous, including: South Bend Art Center; Chicago Public Library; Illinois State Museum; Chicago Botanical Gardens.

Group Exhibitions   Numerous, including: Detroit Institute of Arts, 1955, 1965, 1969; Butler Institute of American Art, 1963-65; Art Institute of Chicago, 1969; Brooklyn Museum, New York, 1972; Smithsonian Institution, 1972; California Palace of the Legion of Honor, San Francisco, 1973; Museum of Fine Arts, Boston, 1975; Miami Art Center, 1975; 28th Annual Boston Printmakers Exhibit, Boston, 1976.

Awards   Numerous, including: Logan Medal, Art Institute of Chicago, 1958; Purchase Prize, University of Minnesota, 1963; Purchase Prize, Ball State University; Exhibition Prize, The Artists Guild of Chicago, 1965; Commonwealth Prize, Detroit Art Institute, 1965; First Prize, 17th Annual Virginia Beach Exhibition, 1972; Purchase Prize, J.B. Speed Museum, 1973; Purchase Prize, Illinois State Museum, 1974; Purchase Award, Fine Arts Center of Indiana, 1975.

Collections   Numerous, including: Illinois State Museum, Springfield; Museum of Fine Arts, Springfield, Massachusetts; Brooklyn Museum of Art, New York; California Palace of the Legion of Honor, San Francisco; Library of Congress, Washington, D.C.; Kalamazoo Institute of Art, Michigan; Virginia Beach Art Center; Art Institute of Chicago; Ringling Museum, Sarasota, Florida; the Universities of Iowa, North Carolina, Southern Illinois, Indiana State, Maine, Michigan State, Nebraska, Utah, Georgia, Louisville, Minnesota, Georgia State.

Works Published in   *Playboy*, 1965, 1967. *North Shore Art League news,* 1969.

254 "Rock Garden XI"
     Color lithograph; 20¾ x 14½"

SEVERA, František

Born   Metánov, Czechoslovakia,
10 January 1924.

Address   Uruguayská 18, 120 00
Prague 2—Vinohrady,
Czechoslovakia.

Education   School of Arts and
Crafts, Prague: diploma.

Career   Free-lance graphic artist,
specializing in scientific and
natural history book illustrations,
1954 to present.

Media   Watercolor, Ink.

Group Exhibitions   Numerous, including the International
Exhibition to Children's Book Illustration, Bologna, Italy,
1977, and shows in Czechoslovakia and other countries.

Awards   3 Gold Medals and 3 awards for illustrations in
books published by Artia.

Works Published in   "Honey producing plants," *Bee-keeping*,
Czechoslovak Agricultural Publishing House, 1960-62.
An atlas of oil-producing plants' diseases and pests.

Czechoslovak Agricultural Publishing House, 1968
[in 4 languages].
Winkler, J. *A pocket guide to beetles.* Prague, Artia, 1969
[in 5 languages].
Kudela, M. *An atlas of pests of coniferous trees.*
Czechoslovak Agricultural Publishing House, 1969.
Richter. *Herbaceous plants.* Czechoslovak Agricultural
Publishing House, 1971.
Vaněk, V. *A concise guide to garden flowers.* Prague, Artia,
1971 [in 5 languages].
Starý, F. and Jirásek, V. *Herbs.* Prague, Artia, 1975
[in 10 languages].
Petrová, E. *A concise guide to flowering bulbs.* Prague,
Artia, 1975 [in 4 languages].
Průcha, J. *A concise guide to flowers from seed.* Prague,
Artia, 1976 [in 4 languages].
Zahrandník, J. *A field guide to insects.* Prague, Artia, 1976
(German), 1977 (English).
Severa, F. (co-author). *European fauna.*
Severa, F. (co-author). *An atlas of cereals.*
An atlas of beet diseases and pests [in progress].
An atlas of pests of deciduous trees [in progress].
Garden in flower [in progress].
A guide to nature [in progress].

255 Pink Hawksbeard, *Crepis rubra*
Watercolor; 9½ x 5¾"

256 Tulip, *Tulipa violacea*
Watercolor; 10 x 6⅜"

257 Blue Lace Flower, *Trachymene coerulea*
Watercolor; 9⅝ x 6½"

258 Obedience, *Physostegia virginiana*
Watercolor; 9½ x 6¾"

259 Crown Anemone, *Anemone coronaria*
Watercolor; 10 x 7¼"

260 Spanish Bluebell, *Endymion hispanicus*
Watercolor; 9⅝ x 6¾"

261 English Yew, *Taxus baccata*
Ink; 9⅜ x 7"

262 Black Nightshade, *Solanum nigrum*
Ink; 9½ x 6⅞"

All lent by the artist

258

SHEEHAN, Marion Ruff (Mrs. Thomas J.)

Born Philadelphia, Pennsylvania, 17 December 1923.

Address 3823 Southwest Third Street, Gainsville, Florida 32607.

Education University of Cincinnati: B.A., Botany, 1946. Cornell University, Ithaca, New York: M.S., Botany and Fine Arts.

Career Assistant Professor of Ornamental Horticulture, University of Florida, 1972 to present (including the teaching of a course in biological illustration). Free-lance botanical illustrator. Formerly, botanical illustrator, L.H. Bailey Hortorium, Cornell University, 1946-52.

Media Ink, watercolor, pastel, gouache, scratchboard, acrylic.

One-person Exhibitions Fairchild Tropical Garden, Miami, 1966; Gainsville, Florida, 1968; the Philippine, Thailand, and Chiengmai Orchid Societies, 1969; 6th World Orchid Conference, Sydney, Australia, 1969; Hunt Institute, 1971.

Group Exhibitions Hunt Institute Internationals, 1964, 1968, 1972; Cleveland Garden Center; Garden Center, Cincinnati; XI International Botanical Congress, Seattle, Washington; International Exhibition of Botanical Art, Johannesburg, 1973; 8th World Orchid Conference, 1975.

Honors/Awards A.O.S. Bronze Certificate, Gainsville Orchid Show, 1971; Honorary Life Member, American Orchid Society, 1974.

Collections Cleveland Garden Center; Public Library, Key West, Florida; University of Florida.

Works Published in Bailey, L.H. *Manual of cultivated plants.* Ed. 2. New York, Macmillan Co., 1949.
Lawrence, G.H.M. *Taxonomy of vascular plants.* New York, Macmillan Co., 1951.
Heller, C.A. *Edible and poisonous plants of Alaska.* Fairbanks, University of Alaska, 1953 (rev. ed. 1958).
Lawrence, G.H.M. *Introduction to plant taxonomy.* New York, Macmillan Co., 1955.
Moore, H.E., Jr. *African violets, gloxinias, and their relatives.* New York, Macmillan Co., 1957.
Noble, M. and Graham, B.H. *You can grow camellias,* Jacksonville, privately published, 1962.
Watkins, J.V. *Your guide to Florida landscape plants.* 2 vols. Gainesville, University of Florida Press, 1963.
Noble, M. Privately published, Jacksonville: *You can garden in Florida,* 1963; *You can grow orchids* (3rd ed.), 1964; *You can grow Cattleya orchids,* 1967; *You can grow Phalaenopsis orchids.* 1971.

263 Lady-Slipper, *Paphiopedilum* sp.
Watercolor; 24¾ x 16¾"
Lent by the artist

264 Ringworm Cassia, *Cassia alata*
Ink; 13 x 9"
Gift of the artist

265 Bull Bay, Southern Magnolia,
Magnolia grandiflora
Ink; 12¾ x 9"
Gift of the artist

263

SINATS, Andrejs

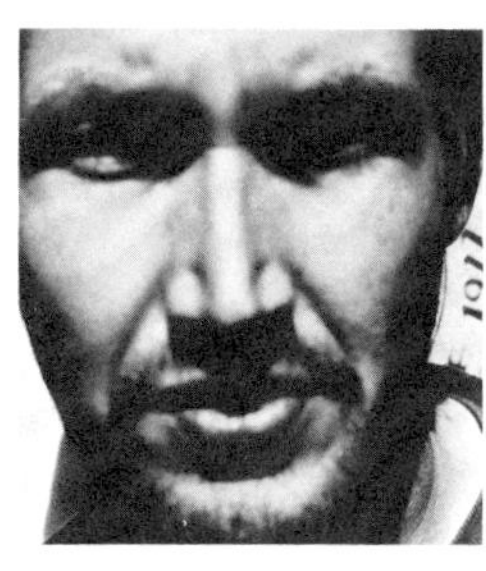

Born   Riga, Latvia, 25 April 1945. United States citizen working in Canada since 1973.

Address   620 Avalon Road, Victoria, British Columbia, Canada.

Education   University of Michigan: B.A., English, 1966. University of Heidelberg: Literature, 1967.

Career   Free-lance printmaker and artist. Produced graphics for Greenpeace Foundation. 1976-77.

Media   Printmaking, painting, photography.

One-person Exhibitions   Mt. Holyoke College, 1972; Malvina Miller Gallery, San Francisco, 1974; Victoria Gallery of Art, Canada, 1974.

Group Exhibitions   Detroit Art Institute travel show, 1976; Minneapolis Institute of Art, 1976; Burnaby Art Gallery, 1976; Victoria Art Gallery, 1976; Western Front, Vancouver, 1977.

Honor   National Endowment for the Arts, 1970.

Collections   More than 20 public collections, including: Albright-Knox Museum, Buffalo; Minneapolis Institute of Art; National Gallery of Canada; Baltimore Museum of Art; Utah Museum of Art; Detroit Art Institute.

266 "Cauliflower"
     Serigraph; 21 x 14¼"

SPOHN, Marvin

Born   Fredonia, Kansas, 24 April 1934.

Died   Los Gatos, California, 21 February 1976.

Education   Washburn University, Topeka, Kansas: B.A., Business Administration, 1956. University of California, Berkeley: Master of Library Science, 1967. Self-taught artist. Studied etching in the studio of Kathan Brown, Berkeley, California, 1969.

Career   United States Army (including 2 years in Germany), 1956-59. Settled in San Francisco, 1960. Accountant with New York Life Insurance Company, 1960-66. Began art career, 1962. Reference librarian and teacher of children's literature, West Valley College, Saratoga, California, 1969-76.

Media   Etching, drypoint, watercolor wash.

One-person Exhibitions   Fredonia Arts Council, Kansas, 1968; Wichita Art Museum, Kansas, 1970.

Group Exhibitions   Numerous, including: Society of Western Artists Annual, M.H. deYoung Memorial Museum, San Francisco, 1967, 1969, 1971; National Academy of Design Annual, New York, 1969; Boston Printmakers Annual, 1970; Mississippi Art Association, 1970; Library of Congress, Washington, D.C.; Smithsonian Institution, Washington, D.C.; California Palace of the Legion of Honor, San Francisco; Crocker Gallery, Sacramento; Oakland Art Museum, California; San Francisco Museum of Art.

Honors/Awards   Several in California, including California State Fair and national competitions. Featured artist by San Francisco Educational Television station KQED, 1974.

Collections   The City of San Francisco; IBM; Wichita Art Museum; University of North Dakota; Castle & Cooke; many other public and private collections. Archival set of his prints deposited at the Achenbach Foundation for Graphic Arts, California Palace of the Legion of Honor, San Francisco.

Works Published in   *Aldebran review,* vol. 8, 1970.
La Revue modern, October 1970, December 1973.
Westart, August 1970.
Occident, fall 1972.
Print editions.
Numerous newspaper reviews and notices.

267 "April Onions," *Allium* sp.
Etching; 12¾ x 11" plate mark

268 "Camellia," *Camellia* sp.
Color etching; 3 x 2⅞" plate mark

269 "Artichoke," *Cynara scolymus*
Color etching; 3⅛ x 5" plate mark

Editions printed by Jeanne Gantz

267

TAGAWA, Bunji

Born   Tokyo, Japan, 13 August 1904. Has been working in the United States since 1922.

Address   22 Veranda Place, Brooklyn, New York 11201.

Education   University of Kansas, Lawrence: A.B., Philosophy, 1926. Cornell University, New York: graduate study in philosophy, 1930.

Career   Free-lance scientific and medical illustrator.

Group Exhibitions   Whitney Museum of Art, New York; Hunterdon Art Center, Clinton, New Jersey, 1977.

Works Published in   Numerous issues of *Scientific American,* 1956-77; and *Hospital Practice,* 1965-74.

270 Amanita Mushrooms: (left to right) *Amanita muscaria;* Destroying Angels, *Amanita phalloides* and *A. virosa; Amanita brunnescens*
Watercolor; 11¼ x 11¼″
Gift of the artist
Cover design, *Scientific American,* March, 1975

271 Agaric Mushroom, *Collybia radicata*
Watercolor; 20¾ x 10½″
Lent by the artist

270

TAN, Yuen Fang

Born   Purwokerto, Dutch East Indies (Indonesia), 14 November 1945. Has worked in the Netherlands since 1969.

Address   Frederikstraat 27 III, 1054 LB Amsterdan, Netherlands.

Education   Gerrit Rietveld Academie, Amsterdam: graphic arts and oils, 1969-71. Studied with Melle, and Charles Jangejans.

Career   Free-lance artist. Formerly, botanical draughtswoman, State Agricultural University, 1971-73. Conducted a course in book illustration, Staatliche Akademie der bildenden Künste, Stuttgart, 1974. Graphic designer, Verlag Kunstkreis Gbbh. & Co., Stuttgart, 1974. Private art instructor and free-lance card designer, 1974-77.

Media   Watercolor, ink, oil, intaglio, woodcut, lithography, pencil.

One-person Exhibition   Galerie Polder, Borne, 1971.

Group Exhibitions   Arti et Amicitae, Amsterdam, 1972; Galerie Polder, Borne, 1974; Museum Fodor, Amsterdam, 1976.

Collections   Numerous public and private collections in Europe.

Works Published in   *Mens en milieu*. Groningen, Wolters-Noordhoff bv., 1971.
van der Maesen, L.F.G. *Cicer L., a monograph of the genus with special reference to the chickpea (Cicer arietinum), its ecology and cultivation*. Wageningen, H. Veenman & Zonen NV., 1972. *
Jaarverslag 1972. Amsterdam, Bührmann-Tetterode NV., 1972.
Breteler, F.J. *The African Dichapetalaceae*. Wageningen, H. Veenman & Zonen NV., 1973. **
Morgen is het jouw beurt. Amsterdam, van Gennep bv., 1973.
Leeuwenberg, A.J.M. "The Loganiaceae of Africa XII, a revision of *Mitreola* L.," Wageningen, H. Veenman & Zonen NV., 1975.
Renaissance (animated film). Amsterdam, K. Schmeink, 1976.

272   Chick-Pea, *Cicer anatolicum* *
Ink; 8¼ x 5½"

273   *Dichapetalum beilschmiedioides* **
Ink; 13⅜ x 9¼"

Permanent loans of the Laboratorium voor Plantensystematiek en -Geografie, Landbouwhogeschool, Wageningen, Netherlands

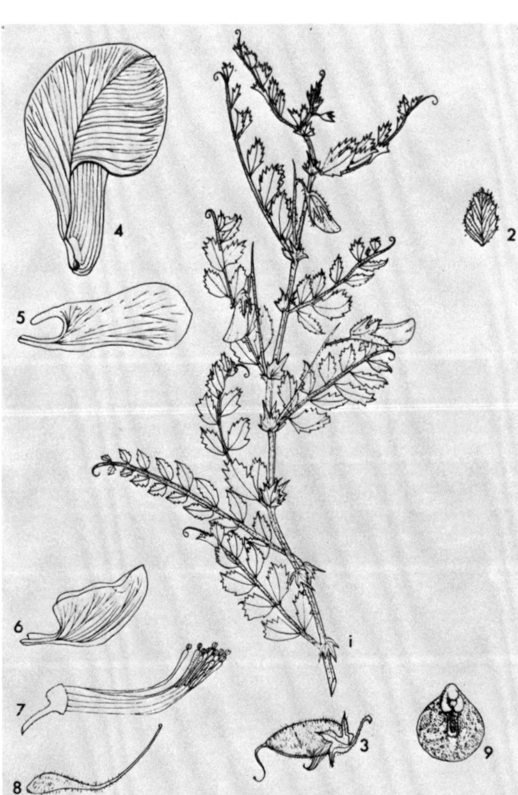

272

TANGERINI, Alice Ruth

Born   Takoma Park, Maryland, 25 April 1949.

Address   2702 South 8th Street, #281-A Arlington, Virginia 22204.

Education   Virginia Commonwealth University, Richmond, Virginia: B.F.A., 1972.

Career   Staff Illustrator, Department of Botany, Smithsonian Institution, 1972 to present. Formerly, free-lance illustrator, Department of Botany, Smithsonian Institution, 1969-72. Graphic artist at Cress, Bethesda, Maryland, and at AIR, Kensington, Maryland, 1969-72. Taught classes in scientific illustration, fall of 1976, and botanical illustration, summer of 1977, to Young Associates of Smithsonian Resident Associates Program.

Media   Ink on Cronaflex, pencil, watercolor, oil.

One-person Exhibition   Virginia Commonwealth University, 1972.

Group Exhibitions   Guild of Natural Science Illustrators Exhibitions: Biocommunication '73, Richmond, Virginia; Implementation '74, New Orleans; Biocommunication '76, Las Vegas.

Award   Second Prize, Juried Show, Virginia Commonwealth University.

Collection   Smithsonian Institution.

Commissions/Works Published in   *Smithsonian magazine, Smithsonian contributions to botany, Bulletin of the Torrey Botanical Club, Kew bulletin, Annals of the Missouri Botanical Garden, Rhodora, Biotropica, Taxon, Baileya, Brittonia, Flora illustrada catarinense, Phytologia, American fern journal, The Bryologist, Science, Journal of the Arnold Arboretum, Oecologia, The United States in the world.*
Ripley, S.D. *Rails of the world.*
Tippo, O. and Stern, W.L. *Humanistic botany.* New York, W.W. Norton, 1977.
Biosystematics and agriculture. United States Department of Agriculture, 2nd Symposium on Systematics [in progress].
Mabrey and others. *Creosote bush.* US/IBP synthesis series.
Poster for Marine Mammal Salvage Program.
Watercolor painting of a Spiny Babbler (bird) for S. Dillon Ripley.

274 *Vellozia hatschbachii*
   Ink on acetate; 12½ x 9¾″

275 *Chusquea scabra*
   Ink on Cronaflex; 17¼ x 14½″
   Gift of the artist

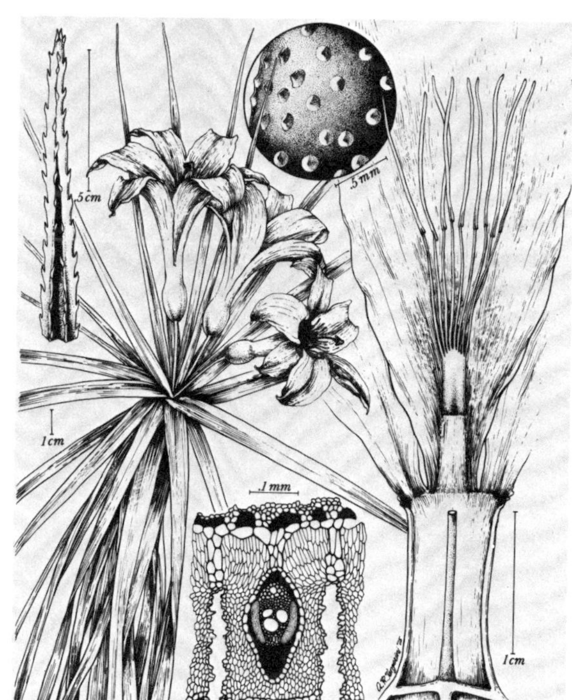

274

TERWINDT-LAGAAY, Marguerite H. F.

Born   The Hague, Netherlands,
17 August 1944.

Address   Beukenweg 12, Velp
(Gelderland), Netherlands.

Education   Academy of Art,
Arnhem: illustration, 1966.

Career   Free-lance illustrator,
specializing in children's books
and fashion illustration.
Formerly, an arts and crafts
teacher. Worked at the State
Agricultural University, 1966-67.

Media   Ink, oil, crayon.

Works Published in   Children's books published by
Callenbach/Nijkerk, and numerous fashion magazines
including *Avenue*.

276 Purple Apricot, *Prunus dasycarpa*
Ink; 9½ x 7⅝″

277 Higan Cherry, *Prunus subhirtella*
Ink; 10¾ x 5¾″

Permanent loans of the Laboratorium voor
Plantensystematiek en -Geografie,
Landbouwhogeschool, Wageningen, Netherlands

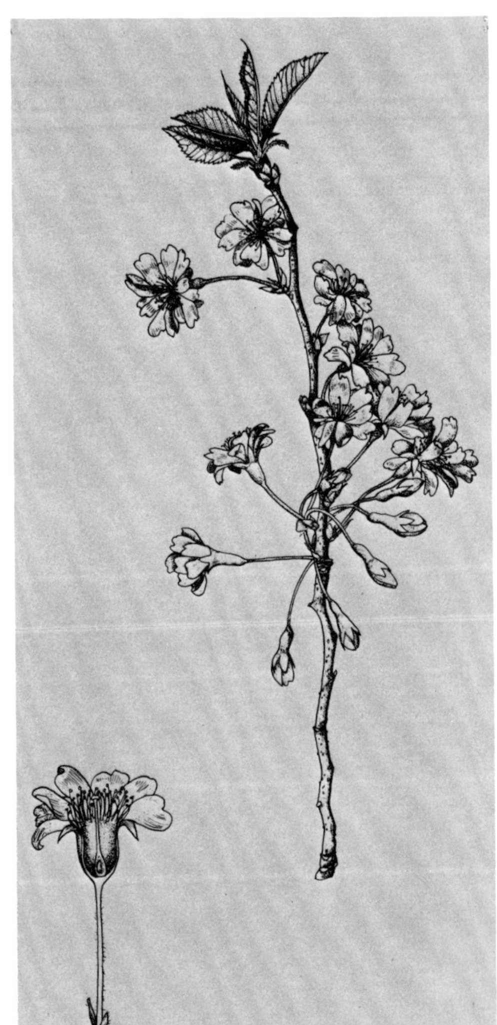

277

TORM, Fernando

Born   Santiago, Chile, 1944.

Address   147 W. 22nd Street,
New York, New York 10011.

Education   Conservatorio Nacional
de Musica, Santiago, Chile. Las
Condes Institute: studied etching
under Rodolfo Opazo. Escuela
Moderna de Musica, Santiago.
Escuala de Belles Artes, Santiago.
The Piano School, New York,
Columbia University, New York.

Career   Printmaker.

Medium   Etching.

One-person Exhibitions   Sala Libertad, Santiago, 1966; Casa
de la Luna Azul, Santiago, 1969; Gallery One,
Poughkeepsie, New York, 1977.

Group Exhibitions   Universidad de Chile, 1964; Carmen
Waugh Gallery, Santiago, 1966; Sausalito, California, 1969;
Valparaiso, 1966; St. Clements Church, New York, 1969;
Galeria Pecanins, Mexico City, 1970; Gallery 100, Princeton,
New Jersey, 1970, 1976; Museo de Arte Moderno, Mexico
City, 1970; State Museum, Trenton, New Jersey, 1970;
Prints Biennial, Santiago, 1970; Toronto, Canada, 1970;
11th Prints Biennial, Ljubljana, Yugoslavia, 1975; Young
Artists Show, New York, 1975; Wavelength, New York, 1976;
Gallery One, Poughkeepsie, New York, 1976; Prints Biennial,
Frechen, Germany, 1976; Museum of Modern Art of Latin
America, Washington, D.C., 1977.

Awards   Scholarship, French Government, Paris, 1962-63.
Scholarship, Pan American Union, New York, 1967-69;
CRAV Award for painting, Santiago, 1966.

278 Bromeliad
Color etching; 16⅝ x 12⅜"

279 Sand Dollar, *Astrophytum asterias*
Color etching; 13⅜ x 11"

280 Cactus hybrid
Color etching; 13¾ x 12½"

278

TRECHSLIN, Anne Marie

Born   Milan, Italy, 17 July 1927.

Address   Huberstrasse 2,
3000 Berne, Switzerland.

Education   Art School, Berne.

Career   Free-lance botanical artist
specializing in rose portraits.

Medium   Watercolor.

One-person Exhibitions   American
Rose Society, Nashville, 1963;
Rome Garden Club, Palazzo
Ruggeri, Rome, 1968; Florence Garden Club, Palazzo Pitti,
Florence, 1969; Hilton Hotel, Malta, 1971.

Group Exhibitions   Hunt Institute Internationals, 1968, 1972;
Hunt Institute Travel Shows (Roses).

Honors   Universal Rose Selection, Meilland named a new
hybrid tea rose after Miss Trechslin, 1968; honorary citizen
of Nashville, Tennessee, 1963.

Collection   National Library, Berne.

Commission/Works Published in   Bois, E. and Trechslin, A.M.
Roses. Edinburgh, Nelson, 1962.
Leroy, A. and Trechslin, A.M. *Les plus belles fleurs de nos
jardins.* Zurich, Editions Silva, 1964.
Leroy, A. and Trechslin, A.M. *Roses.* Zurich, Editions
Silva, Vol. 2, 1967.
Coggiatti, S. and Trechslin, A.M. *Fiori dei nostri giardini.*
Zurich, Editions Silva: II, 1970; III, 1972.
Coggiatti, S. and Trechslin, A.M. *La rosa nel giardino.*
Rome, Ed. R.E.D.A., 1974.
Coggiatti, S. and Trechslin, A.M. *Old garden roses.*
Berne, Ed. Le Moulin, 1975.
Coggiatti, S. and Trechslin, A.M. *Terrazze e giardini.*
Milan, Ed. Rusconi, 1976. *
Ten flower stamps (1970), ten fruit stamps (1971), and ten
bird and flower stamps (1972) for the Republic of
San Marino.
Four rose stamps (1972) and four old rose stamps (1977) for
the Pro Juventute, Switzerland.
Watercolor paintings of roses for Queen Fabiola of Belgium
(1970), Queen Mother Elizabeth of the United Kingdom
(1976), and Queen Sophie of Spain (1977).

281 Common Camellia, *Camellia japonica*
Watercolor; 12 x 8¾"

282 Zinnia hybrid
Watercolor; 9½ x 6"

283 Great Lobelia, *Lobelia siphilitica*
Watercolor; 13¼ x 6½"

284 Day-Lily, *Hemerocallis earliana*
Watercolor; 12½ x 9⅜"

285 Stemless Carline Thistle, *Carlina acaulis*
Watercolor; 7 x 5¼"

286 Chinese Trumpet Creeper, *Campsis grandiflora*
'Tagliabuana' *
Watercolor; 12 x 9"

Gifts of the artist

282

TURKOVIĆ, Greta

Born   Kriẑevci, SRH, Yugoslavia, 11 May 1896.

Address   Ilirski trg 2, Zagreb, SRH, Yugoslavia.

Education   Academy of Arts, Zagreb: A.M., Sculpture, 1925.

Career   Free-lance sculptor, ethnographic and botanical illustrator, and applied artist, 1920 to present.

Media   Watercolor, tempera, ink.

One-person Exhibitions   Split, 1936; Paris, 1937; London, 1967.

Group Exhibitions   Zagreb, 1926, 1935, 1955-56, 1958-59, 1972; Belgrade, 1956; Hunt Institute International, 1972.

Honors/Awards   Diploma, International Exhibition, Paris, 1937; Prize and Diploma, Paris, 1957, praising the *Ampeligraphic atlas;* Silver Grenfell Medal, Royal Horticultural Society, 1967; several awards for applied arts.

Collections   Museum of Applied Arts, Zagreb; National Museum, Split; The Royal Library, Brussels; University of California, Davis; Institut Wadenswil, Switzerland; Academy of Art and Science, Zagreb; Institute for Fruit and Viticulture, Zagreb.

Works Published in   *Dalmatinska ampelografija (Dalmatian ampelography).* Zagreb, P.N. Zavod, 1949.
Turković, Z. and G. *Ampelografski atlas (Ampelographic atlas).* Zagreb, Znanje Press, Part I, 1952; Part II, 1963.
Ljekovito bilje Jugoslavije (Medicinal plants of Yugoslavia). Zagreb, P.N. Zavod, 1960.
Atlas bolesti i štetnika bilja (Atlas of plant diseases and pests). Zagreb, Zadruzna Štampa, 1961.
Jugoslavenska pomologija (Yugoslav pomology). Belgrade, Zadružna Knjiga, 1963.
Hornickel, E. *Der Weinkenner (The wine connoisseur).* Stuttgart, Seewald Verlag, 1965.
Gelenčir, N. *Prirodno liječenje biljem (Healing with plants).* Zagreb, Znanje Press, 1970.
Frazinić, N. *Suvremeno vinogradarstvo (Modern viticulture).* Zagreb, Institute of Viticulture, 1972.
Licul, R. *Praktično vinogradarstvo (Practical viticulture).* Zagreb, Znanje Press, 1972.
Gušić. *Monograph of ancient Croation head-dresses.* 1975.
Book of herbs [in progress].

287   "Waelsh Riesling White" Wine grape,
Vitis vinifera cv.
Watercolor; 14½ x 9⅜″
Lent by the artist

288   Yellow Horned-Poppy, *Glaucium flavum*
Watercolor; 10¾ x 9″

287

TYZNIK, Anthony

Born   Stanley, Wisconsin,
16 November 1925.

Address   Morton Arboretum,
Lisle, Illinois 60532.

Education   University of Wisconsin,
Madison: B.S., Landscape
Architecture, 1950.

Career   Superintendent, Morton
Arboretum, 1953 to present.
Designer and supervisor of many
Morton Arboretum garden entries
in the Chicago World Flower and Garden Shows. Free-lance
landscape architect. Botanical artist. Formerly, worked with
the Milwaukee Park District, prior to 1953.

Media   Ink, pencil, charcoal, oil, watercolor, pastel.

One-person Exhibition   Morton Arboretum.·

Group Exhibitions   Fermi Laboratory, Batavia, Illinois; Hunt
Institute International, 1972; Hunt Institute Travel Shows
(International).

Award   The Hutchinson Medal, Chicago Horticultural Society.

Collection   Morton Arboretum.

Works Published in   *Morton Arboretum quarterly.*

289 Black Walnut, *Juglans nigra*＊
Ink; 14½ x 27½"
Lent by The Morton Arboretum

URBAN, Ladislav

Born   Prague, Czechoslovakia,
26 June 1906.

Address   Na výšinach 16, 170 00
Prague 7, Czechoslovakia.

Education   Classical College,
Prague: degree in painting and
industrial design, 1927.

Career   Free-lance botanical artist
and illustrator.

Media   Watercolor, pencil.

One-person Exhibition   Kostelec nad Černými lesy, 1963
(Watercolors of fungi).

Works Published in   Horynová-Burka. *Zahradnická botanika
(Botany for gardeners)*. 1st edition, 1960; 2nd edition,
1962.
Felix and others. *Květinářství (Cultivation of flowers)*.
Prague, 1963.
Horynová-Macek-Vodáková. *Rostlinná výroba (Vegetable
production)*. 1965.
Lill, K. and others. *Zelinářství (Green grocery)*. Prague, 1971.
Příhoda. *Houbařův rok (Mushroomer's year)*. Prague, 1972.
Příhoda. *Léčivé rostliny (Medical plants)*. Prague, 1973.
Volf and others. *Kvetinárství (Cultivation of flowers)*.
Bratislava, 1976.
"Živa," 1958-76.
Příhoda. *Atlas hub (Book of mushrooms)* [in progress].

290 Spring Snowflake, *Leucojum vernum* (top left);
Winter Aconite, *Eranthis hyemalis* (top right);
English Daisy, *Bellis perennis* (bottom left);
Snowdrop, *Galanthus nivalis* (center);
Coltsfoot, *Tussilago farfara* (bottom right)
Watercolor; 10⅜ x 7″

291 Mezereon, *Daphne mezereum* (top left);
European Filbert, *Corylus avellana* (top right);
Black Alder, *Alnus glutinosa* (bottom left);
Cornelian Cherry, *Cornus mas* (bottom right)
Watercolor; 10⅜ x 7″

292 Medlar, *Mespilus germanica*
Watercolor; 11⅜ x 8¼″

293 Mushroom, *Telamonia armillata*
Watercolor; 9⅜ x 7″

292

VAN DER RIET, Louis Thérèsa Marie Eugenie (Mrs. Paul Maas)

Born   Roosendaal, Netherlands, 24 June 1939.

Address   Vicarius van Alphenlaan 3, Boxtel, Netherlands.

Education   Academy of Arts, Tilburg.

Career   Botanical illustrator, State Agricultural University, Wageningen, 1963-69.

Media   Ink, watercolor, tapestry.

Group Exhibitions   Meermanno Museum, The Hague; Art Centre "Zwaluwenburg," Oldenbroek.

Collection   State Agricultural University, Wageningen.

Works Published in   Breteler, F.J. *The African Dichapetalaceae*. Wageningen, H. Veenman & Zonen NV, 1973.
Voorhoeve, A.G. *Liberian high forest trees*.
Abdallah, M.S. *The Resedaceae*.
Leeuwenberg, A.J.M. "The Loganiaceae of Africa XIV. A revision of *Nuxia* Lam."

294 African Mahogany, *Khaya anthotheca*
     Ink; 13¼ x 9½″

295 *Caylusea latifolia*
     Ink; 10 x 7¼″

Permanent loans of the Laboratorium voor Plantensystematiek en -Geografie, Landbouwhogeschool, Wageningen, Netherlands

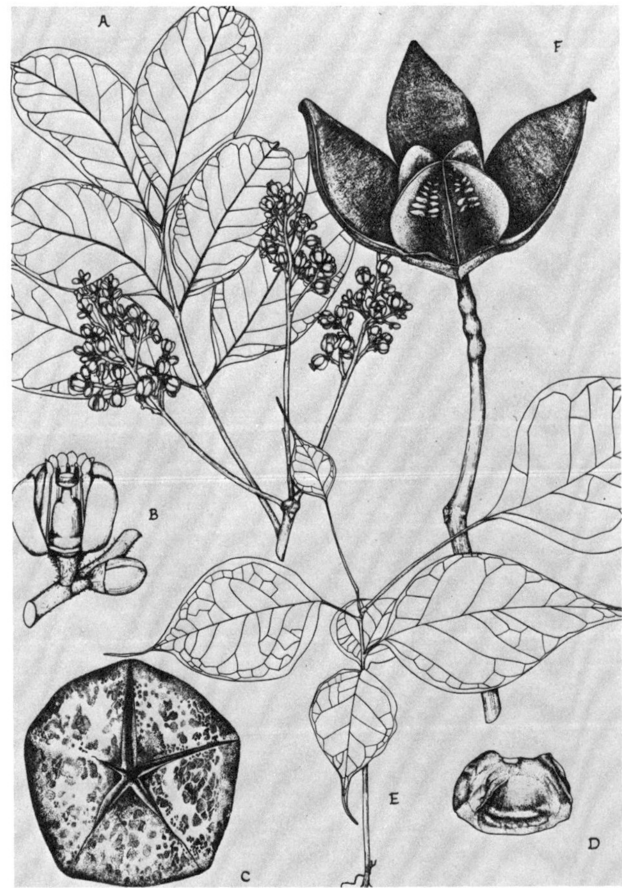

294

VAN TONGEREN, Barend Johannes

Born Utrecht, Netherlands, 12 April 1892.

Died Bennekom, Netherlands, 4 June 1970.

Education Burgeravondschool, Utrecht: 5 year course in lithography, etching, and drawing. Genootschap Kunstliefde, Utrecht. Academy of Arts, Amsterdam: drawing certificate. Studied painting under H.A. van der Wal.

Career Book illustrator. Botanical illustrator at the Universities of Utrecht and Amsterdam. Botanical illustrator, State Agricultural University, Wageningen, 1919-57. Free-lance designer. Still-life, portrait, and landscape painter.

Media Lithography, ink, watercolor, oil.

One-person Exhibition Kunstzaal Eekman, Velp, 1947.

Group Exhibitions De Onafhankelijken exhibitions, Amsterdam, 1916-18; several shows at Wageningen.

Collections State Agricultural University; many private collections.

Commissions/Works Published in Smith, W.S. "Een onderzoek naar het voorkomen en de oorzaken van de verschijnselen, welke worden aangeduid met de naam *ontginningsziekte*," Wageningen, 1927.
Meulen, A. van der. "Over de bouw en de periodieke ontwikkeling der bloemknoppen bij Coffea-soorten," Wageningen, 1939.
Reinders, E. *Histologie en anatomie*. Wageningen, 1943.
More than 13 botanical publications of Landbouwhogeschool, Wageningen, 1920-52.
More than 11 botanical publications of Koninklijke Nederlandse Akademie van Wetenschappen, Amsterdam, 1925-51.
Designed charter for the occasion of the 50 years' Jubilee of H.M. Queen Wilhelmina of the Netherlands.
Designed wedding album for Princess Juliana and Prince Bernard.

296 Nanking or Hansen's Bush Cherry,
Prunus tomentosa
Ink; 5 x 6⅞″

297 Cherry, *Prunus incisa*
Ink; 15¾ x 10¾″

Permanent loans of the Laboratorium voor Plantensystematiek en -Geografie, Landbouwhogeschool, Wageningen, Netherlands

297

VERHEIJ-HAYES, Pamela

Born   Bristol, England, 9 June
1934.

Address   M.A.T.I., Tengeru,
P.O. Box 3101, Arusha, Tanzania.

Education   Swansea High School,
South Wales: General Certificate
of Education, 1948-52. Wye
College, University of London:
1953-57. P.B.N.A., Netherlands:
diploma, garden design, 1965.

Career   Free-lance illustrator
specializing in botany, horticulture, and garden design.
Formerly, botanical illustrator, State Agricultural University,
Wageningen, 1970-71.

Medium   Ink.

Works Published in   Boom, B.K. *Flora der gekweekie
kruidachtige gewassen*. Wageningen, H. Veenman & Zonen,
1970.
Westphal, E. *Pulses in Ethopia, their taxonomy and
agricultural significance*. Wageningen,
Landbouwhogeschool, 1974.

298 Chick-Pea, *Cicer montbretti*
Ink; 8 x 5½"

299 *Dichapetalum brazzae*
Ink; 13⅜ x 9¼"

Permanent loans of the Laboratorium voor
Plantensystematiek en -Geografie,
Landbouwhogeschool, Wageningen, Netherlands

VOGEL, Donella Reese

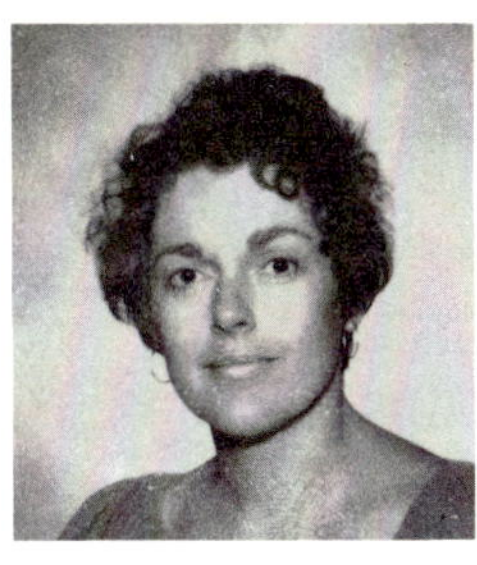

Born   Milwaukee, Wisconsin, 17 October 1942.

Address   8104 East Jefferson Street, Apt. 712C, Detroit, Michigan 48214.

Education   University of Michigan, Ann Arbor: B.S., Design, 1964.

Career   Artist. Formerly, co-owner of Klein-Vogel Gallery, Royal Oak, Michigan, 1970-76. Faculty member, Birmingham-Bloomfield Art Association, Birmingham, Michigan, 1974-76.

Media   Woodcut, intaglio.

Group Exhibitions   Wayne State University Graduate Exhibition, Detroit, 1971; Klein-Vogel Gallery, 1972. Birmingham-Bloomfield Art Association, 1974; Davidson National Print Exhibition, 1975; Birmingham Society of Women Painters, 1976; Michigan Association of Printmakers, 1977.

Award   Purchase Award, Wayne State University, Detroit, 1971.

Collections   In Michigan: Kresge Corporation; City National Bank; Michigan Bell Telephone Company; First Federal Savings and Loan; Bon Secours Hospital, Grosse Point; Manufacturer's National Bank; Beloit Clinic; ACCO Chain-Conveyor Company; Wayne State University. Milton Shoe Company, Milton, Pennsylvania.

Works Published in   *Observer-eccentric newspaper,* Birmingham, Michigan.

300 "Greenhouse"
Etching and aquatint; 24¼ x 18⅜" plate mark

301 "Plants"
Etching; 18⅞ x 18⅞" plate mark

301

WARD, Virginia

Born   Waverly, New York, 23 February 1917.

Address   121 Lincoln Avenue, Albany, New York 12206.

Education   University of Albany: night courses, 1955-60. Studied graphics under Alice Pauline Schafer: 1964-68. Studied pastel under Pauline Edson.

Career   Free-lance artist and private nurse. Holds memberships in: The Print Club of Albany (vice-president); Southern Vermont Art Center, Manchester; Miniature Painters, Sculptors, and Engravers Society, Washington, D.C.; Chaffee Art Gallery, Rutland, Vermont; Catherine Lorillard Wolfe Art Club, New York; National Association of Women Artists, New York; Miniature Art Society of New Jersey, Nutley; Cooperstown Art Association, New York.

Media   Linocut, wood engraving, woodcut, watercolor, oil, pastel.

One-person Exhibition   Chaffee Art Gallery, Rutland, Vermont 1976.

Group Exhibitions   Numerous, including: National Arts Club of New York; The Print Club of Albany; Southern Vermont Art Center; Rensselaer County Council for the Arts, New York; Cooperstown Art Association, New York; Hunt Institute International, 1972; Hunt Institute Travel Shows (International); Third International Graphics Society Exhibition, Hollis, New Hampshire, 1975; National Association of Women Artists Graphics Exhibition, New York and other states, 1975-77; Grist Mill Gallery, Chester Depot, Vermont, 1975; Catherine Lorillard Wolfe Art Club, New York.

Awards   Third Prize for Graphics, Rensselaer County Council for the Arts; 2 Purchase Prizes, The Print Club of Albany; Bronze Medal for Graphics, Catherine Lorillard Wolfe Art Club, 1972.

Collections   The Print Club of Albany; Southern Vermont Art Center; many private collections.

Works Published in   *La revue moderne des arts et de la vie,* September, 1969.

302   "Southeastern Gem," *Billbergia pyramidalis*
Color linocut; 13⅞ x 13⅜"

303   Horse-chestnut, *Aesculus hippocastanum*
Color linocut; 8 x 9"

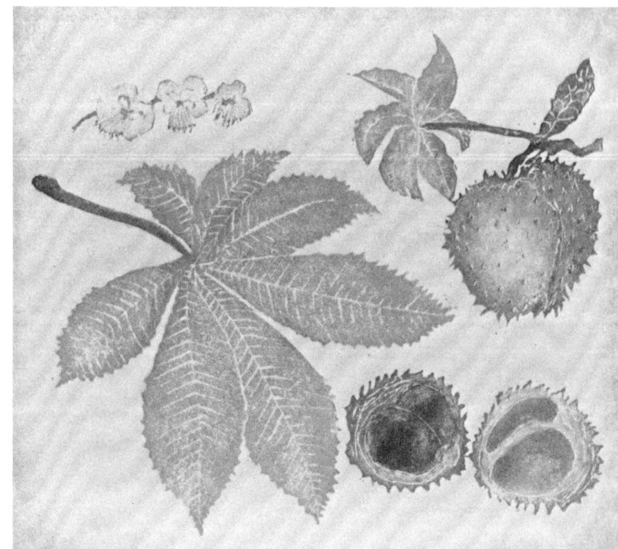

303

WARD-HILHORST, Ellaphie (Mrs. A. W. Ward)

Born   Pretoria, South Africa, 10 July 1920.

Address   5 Tudor Gardens, Pine Road, Kenilworth, Cape Town, South Africa.

Education   Self-taught artist. Studied watercolor drawing with G.P.L. Hilhorst, Netherlands, 1947.

Career   Free-lance artist and illustrator, 1973 to the present. Formerly, mapmaker, 1939-46. Commercial artist, 1946-73.

Media   Watercolor, pencil.

One-person Exhibitions   National Botanic Gardens, Kirstenbosch, Cape Town, 1977; Stellenbosch University, Stellenbosch, 1977.

Collections   Agricultural Technical Services, Pretoria; Stellenbosch University, Stellenbosch.

Works Published in   *Boekcier.* Netherlands, 1950.
Bishop, P.J.G. and Alexander, F.A. *South African bookplates.* Cape Town, A.A. Balkema, 1955.
van der Walt, J.J.A. "The hooded-leaf pelargonium of South Africa," *Veld & flora,* Vol. 62, 1976.
van der Walt, J.J.A. and Ward-Hilhorst, E. *Pelargoniums of Southern Africa.* Cape Town, Purnell & Sons, 1977.*
Dobay, Imre. "Cultivated pelargoniums," *Veld & flora,* Vol. 63, 1977.
Nordenstam, B. "Flowering plants of Africa," *Veld & flora,* Vol. 63, 1977.

304 Geranium, *Pelargonium cucullatum**
Watercolor; 9¾ x 7⅝"

305 Cape Cowslip, *Lachenalia mediana*
Watercolor; 10⅞ x 8¾"

306 Cape Cowslip, *Lachenalia bulbifera*
Watercolor; 12⅜ x 6¾"

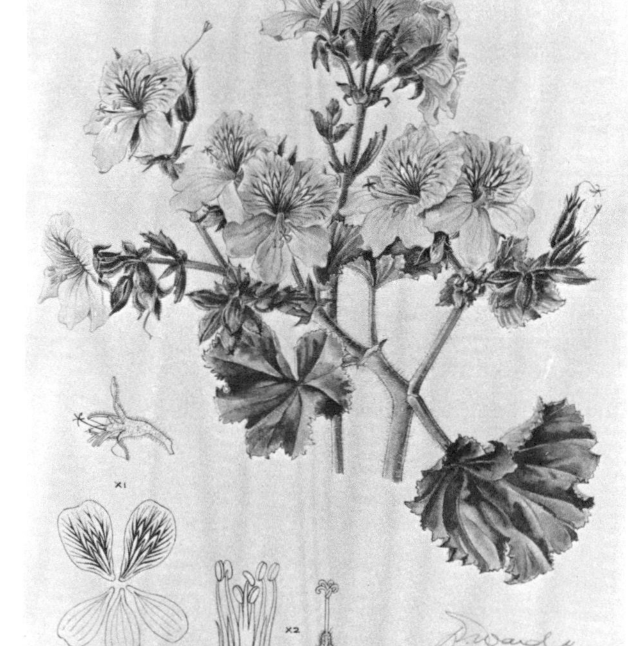

304

WEEMAES, Margot

Born   Louvain, Belgium, 11 June 1909.

Address   Avenue Eudore Pirmez 23, B—1040, Brussels, Belgium.

Education   Higher Institute of Architecture and Decorative Arts of Brussels, Belgian Academy, Rome. Studied at print departments in Brussels, Anvers, London, Rome, Florence, Paris, and Amsterdam.

Career   Free-lance artist and illustrator.

Media   Stained glass, drawing, engraving, lithography.

One-person Exhibitions   Anvers, Bruges, Brussels, Middlekerke, Ostende, The Hague, Dorchester, London, Weymouth, Paris, Rome, Tokyo, Kumamoto, Osaka, etc.

Group Exhibition   Royale Library of Albert I of Belgium, Brussels.

Award   First prize for stained glass, National Basilica of the Sacred Heart, Brussels.

Collections   Cabinet des Estampes, Brussels; Victoria and Albert Museum, London; Museum of Modern Art, Tokyo; Botanical Garden of Brussels.

Commissions   Over fifty stained glass windows in Belgium for the Abbey of Orval; Sainte Elizabeth, Haren; Chapel of Serving Sisters of the Poor, Etterbeek; Sainte Anne, Uccle; Saint Nicholas, Hemiksem.

Works Published in   Publications printed in Anvers by Buschman: *Chansons d'Amures* by Max Elskamp; *Saintes du Brabant; Jeunesse;* and *Proverbs.*
A monograph by Luc Norin [in progress].

307   Hellebore, *Helleborus* sp.
Engraving; 11¾ x 8⅛" plate mark

308   Columbine, *Aquilegia* sp.
Engraving; 5¾ x 3⅜" plate mark
Gift of the artist

309   Rose (Book plate)
Engraving; 3⅞ x 2¾" plate mark
Gift of the artist

310   Bellflower (Book plate)
Engraving; 5⅛ x 3" plate mark
Gift of the artist

307

WESSEL-BRAND, Willy (Mrs. Hugo Wessel)

Born   Arnhem, Netherlands,
21 August 1946.

Address   Paulus Potterhof 23,
Lienden, Netherlands.

Education   Teachers' Training
Schools at Doetinchen, 1968 and
Ede, 1969; Teacher's Certificate in
drawing and manual instruction.

Career   Free-lance artist and
painter. Formerly, botanical
artist, State Agricultural
University, Wageningen, 1968-69.

Medium   Ink.

One-person Exhibition   Tiel, 1972.

Group Exhibition   Museum at Tiel, 1974.

Works Published in   Leeuwenberg, A.J.M. *The Loganiaceae
of Africa VIII Strychnos III.* Wageningen, H. Veenman &
Zonen, 1969.
de Young, P.C. *Flowering and sex expression in Acer L.*
Wageningen, H. Veenman & Zonen, 1976.

311 Apple, *Malus* 'Aldenham Purple'
Ink; 11⅛ x 8″

312 Witch-Hazel, *Hamamelis japonica*
var. *flavopurpurascens*
Ink; 11½ x 7⅝″

Permanent loans of the Laboratorium voor
Plantensystematiek en -Geografie,
Landbouwhogeschool, Wageningen, Netherlands

311

WEST, Keith Robert

Born Amersham,
Buckinghamshire, England,
28 April 1933.

Address Botany Division,
Department of Scientific and
Industrial Research, Private Bag,
Christchurch, New Zealand.

Education Secondary School, High
Wycombe, Buckinghamshire.
Ardmore Teachers Training
College, Ardmore, New Zealand,
1951.

Career Botanical illustrator, DSIR, New Zealand, since
1959. Formerly, art lecturer, Department of Extension
Studies, University of Canterbury, Christchurch, during
1960's.

Media Watercolor, pencil, ink.

Collections Missouri Botanical Garden; DSIR, New Zealand.

Works Published in Fisher, F.F. "The Alpine Ranunculi of
New Zealand," *New Zealand Department of Scientific and
Industrial Research bulletin*, No. 165, 1965.

Sykes, W.R. "Studies of cultivated plants in New Zealand,"
Bignoniaceae, No. 1, 1966.

Mason, R. and West, K.R. "Auckland water plants,"
NZDSIR information series, No. 92, 1973.

Drury, D.G. "Illustrated and annotated key to the erectitoid
senecios in New Zealand," *New Zealand journal of botany*,
No. 12, 1974. *

Healy, A.J. and West, K.R. "Identification of storksbills
(*Erodiuin* ssp.)," *Proceedings of the 27th New Zealand
weed and pest control conference*, 1974.

West, K.R. "Notes on *Epilobium angustum* (Cheesem),"
New Zealand journal of botany, No. 12, 1974.

Healy, A.J. *Identification of weeds and clovers* (second
edition revised). Editorial Services Limited, Wellington,
1976.

Parham, B.E.V. and Healy, A.J. *Common weeds in New
Zealand, an illustrated guide to their identification.*
New Zealand Government Printer, 1976.

Raven, P.H. and T.E. "The genus *Epilobium* in Australasia,"
NZDSIR bulletin, No. 216, 1976.

Wardle, P. *Plants and landscape of Westland National Park*
[in progress].

West, K.R. and Raven, P.H. "Novelties in Australian
Epilobium," *New Zealand journal of botany* [in progress].

313 Willow Herb, *Epilobium brevipes*
Watercolor; 7½ x 6½"

314 Groundsel, *Senecio minimus**
Ink; 10⅝ x 6⅜"

315 Dandelion, *Taraxacum officinale*
Pencil; 10⅛ x 13⅜"

All indefinite loans of the artist

315

WHAITE, Gillian Rosemary

Born   London, England, 15 April 1934.

Address   30 Love Walk, London SE5 8AD, England.

Education   University of London: Slade Diploma of Fine Art, 1955; Art Teacher's Certificate, 1959. Royal Academy Schools, London: sabbatical year 1968-69.

Career   Artist and Senior Lecturer in Art, St. Gabriel's College of Education, London, 1959-78.

Media   Oil, watercolor, ink, graphics (particularly etching).

Group Exhibitions   Royal Academy Summer Exhibitions; New English Art Club; S.E.A.; R.B.A.; and other London exhibitions.

Collections   St. Gabriel's College Collection, London; private collections in England, United States, and Canada.

316 "Lilies," *Lilium speciosum*
    Etching; 24¼ x 18″ plate mark

317 "Flowers in February"
    Etching; 24 x 18¼″ plate mark

316

WYNNE, Patricia J. (Mrs. Maceo Mitchell)

Born   Chicago, Illinois, 17 August 1945.

Address   446 Central Park West.—4E, New York, New York 10025.

Education   Illinois Wesleyan University, Bloomington; B.F.A., Graphics, 1967. University of Iowa, Iowa City: M.F.A., Printmaking, 1970.

Career   Free-lance artist. Formerly, Assistant of Graphics, Illinois Wesleyan University, 1967. Art History Research Assistant, University of Iowa, 1968-70. Instructor of Art History and Design, Macomb County Community College, Warren, Michigan, 1970-71. Instructor of Drawing, Wayne State University, Detroit, 1971-72. Instructor of Prints and Drawing, University of Windsor, Windsor, Ontario, Canada, 1971-73.

Group Exhibitions   Chicago Art Institute, 1966; Evansville, Indiana, 1966; Waukesha, Wisconsin, 1967; McLean County Bank, Bloomington, Illinois, 1967; Moline, Illinois, 1968; Wesley Foundation Center, Iowa City, 1969; Shelter Gallery, Royal Oak, Michigan, 1970; Wayne State University, Detroit, 1971; University of Windsor, 1973; Omaha, Nebraska, 1973; Macomb County Community College, Mt. Clemens, Michigan, 1973; University of Wisconsin, Marshfield, 1973.

Collections   Baxter Medical Center; Beznos Collection; Joanne Marie Davis; Harris Art and Design; McLean County Bank; National Bank of Detroit; Silverman Gallery, Chicago; University of Michigan; University of Wisconsin.

Commissions/Works Published in   *Arkansas Game and Fish commission,* Vol. 7, winter, 1977.
Soucie, G. "Consider the shark," *Audubon,* September, 1976.
Reiger, G. "The white amur caper," *Audubon,* September, 1976.
Wynne, P.J. *The animal abc's.* New York, Random House, 1977.
Cover design for the *Field Museum of Natural History bulletin,* December, 1974.
Illustrations for various articles in *Time-life nature/science annual,* 1976-77.
Various illustrations for the American Museum of Natural History, New York.

318 "Irises," *Iris* sp.
Color pencil; 26 x 32½" sheet size

WYSS-FISCHER, Verena

Born Zürich, Switzerland, 31 August 1942.

Address Unterburg 10, 8158 Regensberg, Switzerland.

Education Kunstgewerbeschule, Zürich: Diploma, Scientific Illustration, 1963.

Career Free-lance scientific illustrator.

Media Watercolor, ink.

Commissions *Ciba-Geigy-Unkrauttafeln*. Basel, Ciba-Geigy AG, 1968-70. Three flower stamps, Pro Juventute series, 1975, 1976, Switzerland.

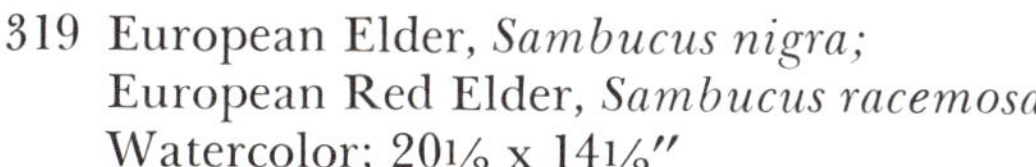

319 European Elder, *Sambucus nigra;*
European Red Elder, *Sambucus racemosa*
Watercolor; 20½ x 14½″

ZEWALD, Henrike Gerardi Dorothea

Born   Buitenzorg, Java, Dutch East Indies (Indonesia), 20 October 1929.

Address   Torckstratt 9, Velp (Gelderland), Netherlands.

Education   Academy of Art, Arnhem: diploma in jewelry design, 1949-54. Self-taught botanical artist.

Career   Botanical artist, State Agricultural University, Wageningen, 1955 to the present.

Medium   Ink.

Group Exhibition   1,000 Years of Flower Illustration, the National Meerman and Westreenen Museum, The Hague, 1966.

Commissions/Works Published in   de Wit, H.C.D. "A revision of *Afrardisia*," *Blumea*, 1958.

Breteler, F.J. "A revision of *Abrus*," *Blumea*, 1960.

Leeuwenberg, A.J.M. "The Loganiaceae of Africa I: *Anthocleista*," *Acta botanica Neerlandica*, 1961; "II: Mosterea," *Mededelingen Landbouwhogeschool*, Wageningen, 1961.

Hugos, T.H. "A revision of the genus *Parkia*," *Acta botanica Neerlandica*, 1962.

Oldeman, R.A.A. "A revision of the genus *Didelotia*," *Blumea* 1964; and *Acta botanica Neerlandica*.

Risseeuw, M. A revision of the genus *Buchholzia*," *Acta botanica Neerlandica*, 1964.

Amshoff, G.J.H. "Myrtacées," *Flore du Gabon*, 1966.

de Wit, H.C.D. *Aquarium planten*. Hollandia-Baarn-Nedeeland, 1966; *Aquarium Pflansen*. Stuttgart, Eugen Ulmer, 1971.

Bos, J.J. "The genus *Liparia*," *Journal of South African botany*, 1967.

de Wilde, J.J.F.E. "A revision of the species of *Trichilia* on the African continent," *Mededelingen Landbouwhogeschool*, Wageningen, 1968.

Leeuwenberg, A.J.M. "The Loganiaceae of Africa VIII: *Strychnos* III," *Mededelingen Landbouwhogeschool*, Wageningen, 1969.

van der Maesen, L.J.G. "The genus *Cadia*," *Acta botanica Neerlandica*, 1970.

van der Maesen, L.J.G. "A monograph of the genus *Cicer*," *Mededelingen Landbouwhogeschool*, Wageningen, 1972.

Breteler, F.J. "The African Dichapetalaceae." *Mededelingen Landbouwhogeschool*, Wageningen, 1973.

Westphal, E. "Pulses in Ethiopia," *Belmontia*, 1974.

Grubben, G.J.H. "The cultivation of *Amaranthus*, a tropical leaf vegetable," *Mededelingen Landbouwhogeschool*, Wageningen, 1975.

Eimunjeze, V.E. "A revision of *Hemandradenia*," *Mededelingen Landbouwhogeschool*, Wageningen, 1976.

de Wilde-Duyfjer, B.E.E. "A revision of the genus *Allium* in Africa," *Mededelingen Landbouwhogeschool*, Wageningen, 1976.

320 Cowpea, *Vigna unguiculata*
   Ink; 8 x 5⅞"

321 *Trichilia emetica*
   Ink; 13¼ x 8¾"

Permanent loans of the Laboratorium voor Plantensystematiek en -Geografie, Landbouwhogeschool, Wageningen, Netherlands

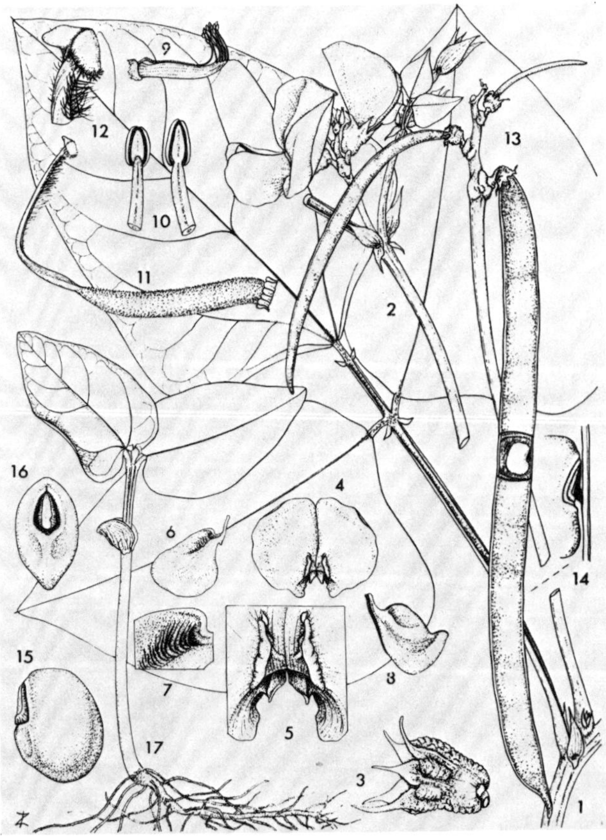

320

ZOCCHIO-FIDALGO, Carmen Sylvia (Mrs. Oswaldo Fidalgo)

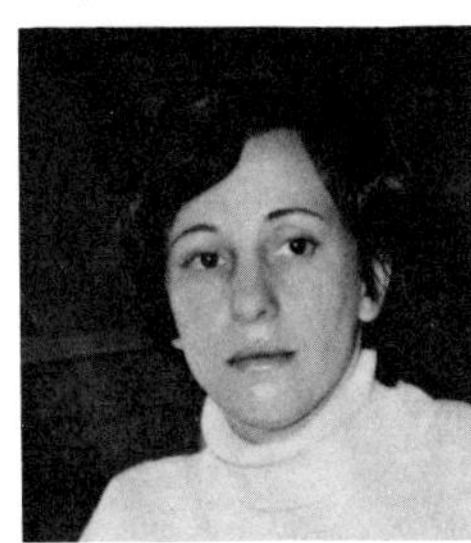

Born   São Paulo, Brazil, 7 August 1941.

Address   Instituto de Botânica, Caixa Postal 4005/01000, São Paulo, SP, Brazil.

Education   Institute of Education "Caetano de Campos": Teacher of Drawing degree, 1962. Studied oil painting and pastel with Colette Pujol, and oil painting with Durval Pereira. "Fundação Armando Alvares Penteado," São Paulo: studied plastic arts, 1976.

Career   Head of the Section of Botanical Illustration, Institute of Botany, São Paulo, Brazil.

Media   Watercolor, oil, ink, pastel.

Group Exhibitions   Hunt Institute International, 1972; Hunt Institute Travel Shows (International); Hunterdon Art Center, Clinton, New Jersey, 1977.

Collections   Institute of Botany, São Paulo; World Life Research Institute, California.

Works Published in   Galli, F. and others. *Manual de fitopatologia*. São Paulo, Editora Agronômica Ceres, 1968. Bicudo, C.E.M. and R.M.T. *Algas de águas continentais Brasileiras* [cover]. São Paulo, Fundação Brasileira para o Desenvolvimento do Ensino de Ciências, 1970. Illustrations in *Rickia, Hoehnea, Nova Hedwigia*.

322 Garland-Flower,
Hedychium coronarium
Watercolor; 22 x 26″

323 Garland-Flower,
Hedychium gardnerianum
Watercolor; 23 x 26½″

Both lent by the artist

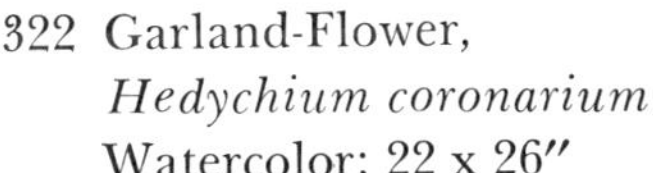

323

Cumulative Artist Index

Contemporary botanical artists and illustrators represented in the 1st, 2nd, 3rd, and/or 4th International Exhibitions of Botanical Art and Illustration (1964, 1968, 1972, 1977) at the Hunt Institute.

Foster, Mulford B. III
Forberg, Janice R. III
Frazier, Vivien L. III, IV
Frederick, Frances D. III
Fujishima, Junzō II
Fukuda, Haruto III, IV
Fukuda, Marlene III
Furse, John P.W. I
Fusi, Almina D. IV

Gadd, Arne W. II
Gardner, Sue IV
Gavriloff, Alejandro III
Gerrard, Primrose III, IV
Ghislain, Raphael H.-C. III, IV
Gilkey, Helen M. II
Golte-Bechtle, Marianne III, IV
Goodall, Natlie P. III
Gordon, John IV
Gossner, Gabriele II
Goulandris, Niki II
Greene, Wilhelmina F. II
Greenwood, Leslie II
Grierson, Mary A. II, III, IV
Griswold, Ralph E. I
Grosch, Laura IV
Grossman, Karl II
Gruter, Adrianus IV
Guinea, Emilo II
Günthart, Lotte II

Hainard, Robert I
Hall, Clarence E. III
Hallé, Francis II
Hallé, Nicholas M.P. II
Haller, Isabelle C. I
Halliday, Patricia III
Hamilton, Lucretia B. III
Hammond, Catherine IV
Hanna, Boyd E. II, III
Hansens, Aline L. II

Harper, Charles B. III
Harrison, Olga F.E. III
Harshbarger, Gretchen F. I
Hart, Nancy S. III, IV
Harvey, Norman B. III, IV
Hasegawa, Kiyoshi II, III
Hassen, Anne I
Hattal, Rita II
Hayes, Marvin IV
Heins, Esther III, IV
Helmer, Jean C. IV
Hennessy, Esmé F.F. IV
Herklots, Geoffrey A.C. IV
Hermes, Gertrude III
Higgs, Helen B. III
Hilhorst, Gerhardus P.L. IV
Hnizdovsky, Jacques III, IV
Hoek, Adriana E.A. IV
Holgate, Jean II, III
Holladay, Harriett I
Holmqvist, Kristian IV
Hood, Mary V. I
Hoss, Della T. III
Hottes, Alfred C. I
Houck, Margaret III
Howe, Marsha K.H. IV
Hughes, Regina O. III
Hull, Gregory IV
Huntley, Victoria H. I
Hutchinson, John II, III

Inagaki, Tomoo III
Irwin, James B. III
Iselin-Gueydan, Sophie III
Ito, Nanae II

Jamison, Philip IV
Janish, Jeanne R. I
Johnson, Arthur M. I
Johnston, Edith F. I, II

Jonzon, Carl V. II
Jules, Mervin II, III, IV

Kadak, Tiiu III
Kamel, Gamil I. III
Kanamori, Yoshio III
Kasubski, Sylvia II
Kawarazaki, Shōdō III
Kelly, Stan III
Kers, Lars E. II
Kimber, Sheila V. II
Kitaoka, Fumio III
Klein, Karen A. IV
Kleinert, Hans III
Knigin, Michael IV
Kobinger, Hanns III
Koch, Ernöne II
Kogan, Jane III
Kohlmeyer, Erika O. III
Korth, Marianne T. III
Koyama, Tetsuo M. III, IV
Krause, Erik H. II, III
Kredel, Fritz I
Krejča, Jindřich III
Kruckeberg, Mareen S. III
Kunz, Otto L. IV

Landacre, Paul H. I
Landwehr, Jacob II
Lanfranco, Guido G. III
Langedijk, Gerrit J. IV
Leak, Virginia L. I
Leake, Dorothy III
Leal, Newton P. II
Leatherwood, Linda B. IV
Leighton, Clare I, II
Lessnick, Howard III
Letty, Cythna I, II
Lexa-Regéczi, Marta II, III, IV
Leyniers de Buyst, Dorika II, III
Lid, Dagny Tande I, III, IV

Lieberman, Frank J. I
Lindström, Bibbe II
Linnamies, Sirkka I. III
Little, Robert W. III
Llanacoplos, Hortensia III
Loewer, Henry P. IV
Long, Lois IV
Lucioni, Luigi III, IV

Mack, Warren B. I
Mackay, Donald A. IV
Mackley, George E. II
Maggiora, Laura R. IV
Mahnke, Heinz K.G. II, III
Maltzman, Stanley IV
Manhã-Ferreira, Hilda II
Marinsky, Harry III
Marlier, Fritz II
Martin, Colette I, II, III
Mastick, Marian F. II
Maury, Anne E. II
McBride, Leslie R. II
McCubbin, Charles W. III, IV
McEwen, Rory I, IV
Meadows, Gillian A. III
Mee, Margaret U. II
Megaw, Elektra II
Mentges, Charlotte I
Mergner, Hans III
Merrilees, Rebecca I
Miller, Grambs III
Möckel, Else III
Mockel, Henry R. I, II, IV
Monsma, Mary S. IV
Moran, Margaret A. IV
Morris, Dennis I. III
Morrison, Benjamin Y. II
Moser, Barry IV
Mourré R., Leonardo III
Murdoch, Florence I
Murman, Eugene O.W. I

Nagase, Fujiko III
Nakayama, Mitsu I
Nanao, Kenjilo IV
Nash, John II, III
Nehring, Harald III
Nesbitt, Lowell B. III
Newsome-Taylor, Cynthia II, III
Newton, Betty J. I
Nicholson, Barbara E. IV
Ničova-Urbanová, Věra IV
Nordenskjöld, Birgitta A.K. II
Noritake, Chizuko III
Norman, Edward d'Aubigny IV
Norman, Marcia G. II, IV
Norrman, Gunnar II

Ogden, Nancy A. III
O'Gorman, Helen I
O'Grady, Mona III
Ohta, Yoai IV
Ospina H., Mariano II

Pahl, Marion IV
Park, Dorothy L. II
Papp, Charles S. I, III
Parham, Margaret E. III
Parker, Agnes M. III
Patten, Gerry V. I
Persson, Karin II
Pertchick, Bernard I
Pertchick, Harriet I
Peterson, Norton III, IV
Pingitore, Eugenio J. IV
Pinheiro, Jane S. I
Pistoia, Marilena IV
Poluzzi, Carlo II, III
Pomeroy, Mary B. I, III, IV
Poole, Monica III, IV
Povall, Lois IV
Powell, Linda K. IV

Prentiss, Thomas S. IV
Priest, Hartwell W. III
Prince, Martha IV
Prins, Johanna M.C. II
Pugh, Edward J. I
Purves, Rodella A. III

Ravindran, O.T. II, III, IV
Ravnik, Vlado III
Raynal, Aline M. II
Readio, Wilfrid A. I
Reid, Emily E. I
Reiner, Imre II
Richards, Albert G. I
Rickett, Harold W. I
Riefel, Carlos II, III
Riemer-Gerhardt, Elisabeth IV
Rissler, Gerd III, IV
Roberts, Mary V. II
Roberts, Patricia H. I
Roca-Garcia, Helen III
Rodriguez, Rafael L. II
Roesener, Richard IV
Rollinat, Madeleine III
Rosario, Arnulfo C. del II
Roshardt, Pia S. II
Ross-Craig, Stella II
Rosser, Celia E. III
Rosvall, Steffan II
Runyan, Frances E. III
Rushmore, Arthur W. II

Saito, Charles M. III, IV
Saldanha, Cecil J. III
Salgado, Eduardo A. III
Salvado, Samuel O. II, III
Sanchez, Emilio III
Sanders, Merritt D. II
Santos, Norga M. IV
Sartain, Emily I
Schafer, Alice P. III, IV

Scheepmaker, W. IV
Schlageter, Laura II
Schroeder, Jack III, IV
Schroer, Elisabeth IV
Schwartz, Carl E. IV
Seagrief, Stanley C. II
Segura, Juana R. III
Seitz, Marta III
Sekino, Junichiro III
Severa, František IV
Sheehan, Marion R. I, II, III, IV
Sherwood, Harold F. I
Sierra-Ràfols, Eugeni II
Silva, Fanny A. III
Sinats, Andrejs IV
Smith, Alice U. III
Smith, Ethelynde I
Smith, Rosemary M. III
Smith, Susan C. II
Snelling, Lilian I, III
Spohn, Marvin E. III, IV
Stones, E. Margaret I, II, III
Svolinský, Karel II

Tagawa, Bunji IV
Takal, Peter II, III
Tan, Yuen Fang IV

Tangerini, Alice IV
Terwindt-Lagaay, Marguerite IV
Threlkeld, Gesina B. II
Tillett, Stephen S. I
Torm, Fernando, IV
Tourjé, Margaret S. I
Trechslin, Anne M. II, III, IV
Tröger, Annemarie II, III
Tschaffon, Wanda M. II
Tseng, Charles C. I
Turković, Greta III, IV
Tyznik, Anthony III, IV

Urban, Ladislav IV
Ursing, Björn III
Usak, Otto II

Van der Riet, L.T.M.E. IV
van Tongeren, B.J. IV
Vasudevan, R. Nair III
Veiss, Maria E. III
Velick, Bernadette S. I
Verheij-Hayes, Pamela IV
Vodičková, Vlasta III
Vogel, Donella IV
Volpe, Richard III
von Below, Irma II

Ward, Minor F. III
Ward, Virginia III, IV
Ward-Hilhorst, Ellaphie IV
Watling, Roy III
Watson, Patricia I
Weatherwax, Paul III
Webber, Irma E. III
Weemaes, Margot IV
Weisz, Josef II
Werneck de Castro, Maria II
Wessel-Brand, Willy IV
West, Keith R. IV
Whaite, Gillian R. IV
Wheeler, Zona L. II
Williamson, Gene III
Wood, Heather III
Wynne, Patricia IV
Wyss-Fischer, Verena IV

Yoshida, Toshi III

Zahn, Martin II
Zapf, Hermann II
Zetterholm, Torbjörn II, III
Zewald, Henrike G.D. IV
Zocchio-Fidalgo, Carmen S. III, IV
Zwinger, Ann H. III

❧ Credits ❧

Portraits R. Alava: Marion Cave, 1959. D. Allen: Walter Hodge. J. Andrews: *East Anglian Daily Times* and Associated Papers. M. Angel: G.C. Newman. G. Barbisan: Galleria d'Arte San Giorgio, Mestre, 1976. S. Beckett: Self-portrait. R. Bero: Associated American Artists. M. Blos: Peter Blos. H. Borkowski-Braendlin: Ruth Vögtlin. A. Cleuter: *La Lanterne,* Brussels. A.O. Dowden: Matthew Wysocki. H. Evans: Marsha Evans. P. Fawcett: R. Pardo. E. Felsko: Wilhelm Preim. S. Gardner: C.W. Carpenter. P. Gerrard: James H. Gerrard, Jr. L. Grosch: Earl Lawrimore. N. Hart: William S. Stickney. N.B. Harvey: Rees Osborne. M. Hayes: A. W. Merwin. E. Heins: Rick Stafford. E.F. Hennessy: Mrs. E. Lorraine Mulder. J. Hnizdovsky: Self-portrait. A.E. Hoek: M. Hoek. M. Howe: Looart Press. G. Hull: Yasuo Minegawa. K. Klein: Richmond Rounds. L. Leatherwood: Robert Palmer Photographics. D.T. Lid: Lone Teilmann. H.P. Loewer: Harold Maas. L. Long: Yasuo Minegawa. L. Lucioni: Einars J. Mengis, Shelburne Museum, Inc. D. Mackay: John Bickel. C. McCubbin: John Borthwick. B. Moser: Unitas. K. Nanao: Victor Wong. B. Nicholson: Ashoke Pasi. E. and M. Norman: Herbert P. Vose. M. Pomeroy: Jim Ziegler. G. Rissler: Aftonbladet, Stockholm. R. Roesener: Richard Roesener. C. Saito: Frame House Gallery, Inc., Louisville, Kentucky. N.M.M. dos Santos: Instituto de Botânica, São Paulo, Brazil. J. Schroeder: Austin Walmsley. C. Schwartz: Les Klug. B. Tagawa: Ikuyo Garber. A. Tangerini: Peter Harholdt. L. Urban: Jiří Haager, 1974. C. Zocchio-Fidalgo: Instituto de Botânica, São Paulo, Brazil.

The following exhibition items were purchased with the aid of a grant from the National Endowment for the Arts, Washington, D.C.: Numbers 1, 2, 30, 45, 74-77, 91, 94, 98, 99, 108, 109, 123, 140, 141, 145, 146, 167, 172-175, 178, 179, 185, 186, 213, 218-220, 231, 241-243, 246-248, 251, 254, 267-269, 274, 298-301, 316.

Catalogue Design: Sally Secrist   Printing: Pickwick-Morcraft, Inc., Pittsburgh